Green Energy and Technology

Climate change, environmental impact and the limited natural resources urge scientific research and novel technical solutions. The monograph series Green Energy and Technology serves as a publishing platform for scientific and technological approaches to “green”—i.e. environmentally friendly and sustainable—technologies. While a focus lies on energy and power supply, it also covers "green" solutions in industrial engineering and engineering design. Green Energy and Technology addresses researchers, advanced students, technical consultants as well as decision makers in industries and politics. Hence, the level of presentation spans from instructional to highly technical.

Indexed in Scopus.

Indexed in Ei Compendex.

Marta Torres González · Carlos Rubio Bellido
Editors

Towards Low and Positive Energy Buildings

Thermal Comfort, Climate Change, and Energy Efficiency Approaches

Springer

Editors
Marta Torres González
Building Construction II
University of Seville
Seville, Spain

Carlos Rubio Bellido
Building Construction II
University of Seville
Seville, Spain

ISSN 1865-3529 ISSN 1865-3537 (electronic)
Green Energy and Technology
ISBN 978-3-031-70850-3 ISBN 978-3-031-70851-0 (eBook)
https://doi.org/10.1007/978-3-031-70851-0

This Springer imprint is published by the registered company Springer Nature Switzerland AG
The registered company address is: Gewerbestrasse 11, 6330 Cham, Switzerland

Prologue

In an era where the impacts of climate change are increasingly evident, addressing the interconnected challenges of thermal comfort and energy production has never been more critical. This book brings together the insights and expertise of scientific researchers from around the world, each contributing their unique perspectives and findings to a collective effort aimed at understanding and mitigating these global issues.

This book is part of the results of the ENERPOT+ project, Positive Energy Buildings Potential for Climate Change Adaptation and Energy Poverty Mitigation (PID2021-122437OA-I00), funded by the Spanish Ministry of Science and Innovation and directed by Professor Carlos Rubio Bellido of the University of Seville. This work has been developed within the Ibero-American network on energy poverty and environmental well-being (722RT0135) as part of the Ibero-American program of science and technology for development lead by the Professor Alexis Perez Fargallo.

We extend our deepest gratitude to all the authors who have contributed to this volume, whose dedication and expertise have made this publication possible. We also wish to thank all the collaborators whose support and hard work were invaluable in bringing this project to fruition.

Divided into three comprehensive parts, this volume explores the multifaceted nature of climate change, its implications for thermal comfort, and the vital role of energy production in both exacerbating and alleviating energy poverty. Each chapter delves deeply into specific aspects of these themes, presenting cutting-edge research and innovative solutions that are essential for policymakers, industry leaders, and scholars alike.

Part One: Energy Efficiency and Building Performance examines the intersection of energy efficiency and building performance through various case studies and evaluations. From technical-economic assessments to innovative energy management strategies, this part provides insights into enhancing building sustainability.

Part Two: Energy Performance Assessment and Thermal Comfort focuses on assessing energy performance and enhancing thermal comfort in diverse contexts. Case studies range from mapping energy performance in residential buildings to studying thermal comfort in tropical and Mediterranean climates, offering valuable insights for sustainable building practices.

Part Three: Sustainable Solutions and Climate Adaptation explores sustainable solutions and climate adaptation strategies. From integrating solar PV in Mediterranean buildings to assessing urban heat island effects and sustainable urban flooding solutions, this section emphasizes practical approaches to addressing climate challenges.

The collective work presented in these chapters underscores a common goal: to foster a deeper understanding of how our actions today will shape the future of our planet. The authors, hailing from Spain, Japan, Portugal, Mexico, Chile, Ecuador, Costa Rica, and Uruguay, bring a rich diversity of knowledge and experience, reinforcing the notion that collaborative efforts are essential in tackling the complex challenges posed by climate change.

We hope that this publication will prove highly useful for construction professionals, researchers, innovators, and all those interested in a deeper exploration of these critical issues. As you delve into the following chapters, we invite you to consider the profound implications of the research presented and to join us in envisioning and striving for a sustainable, comfortable, and equitable world.

School of Building Engineering, Seville (Spain)

Marta Torres González
Assistant Professor and Researcher
University of Seville
Seville, Spain

Carlos Rubio Bellido
Associate Professor and Vice-dean for
Mobility and International Relations
University of Seville
Seville, Spain

Contents

Energy Efficiency and Building Performance

Greening the Deal: Climate Change Adaption Through Circularity and Energy Efficiency in Buildings

Maria Paula Mendes and Inês Flores Colen

Abstract Europe has undergone rapid warming in recent years, resulting in frequent and intense heat waves during the past two decades. This warming trend in Europe has resulted in prolonged periods of extreme heat and amplified the atmosphere's ability to retain water vapor. Consequently, the growing capability of moisture retention leads to more intense rainfall and storms, increasing the likelihood of more frequent flood events. These interlinked events shed light on the intricate relationship between human activities, the emission of greenhouse gases (GHG), and the dynamics of climate change, revealing its complex nature. Considering these developments, the costs associated with adapting various structures, ranging from residential homes to educational institutions and healthcare facilities, alongside fortifying flood defenses, exhibit an upward trajectory across all climate scenarios when contrasted with scenarios void of climate change impacts (with high confidence). Additionally, this accelerated warming trend has implications for the durability of materials such as wood, plastic, stone, paint, and renders, which are susceptible to degradation caused by solar radiation and hygrothermal conditions. Circular economy and energy efficiency in buildings are essential for achieving sustainability and climate mitigation. The circular economy model promotes a more responsible use of natural resources and their associated environmental impacts, with long-term strategies that preserve buildings and formulate resilient solutions with proper maintenance. Likewise, energy efficiency in the construction sector can reduce energy consumption and emissions, achieving a good balance between passive and active solutions. For example, in the EU-27, in 2021, 27% of GHG emissions came from the energy sector, 23% from industry, 14% from the residential and commercial sectors, and 3% from waste. This chapter explains the impact of climate change in Europe and the potential effects on the built environment, demonstrating how circularity and energy efficiency in buildings serve as strategies for achieving climate change

M. P. Mendes (✉)
CERENA, Associação do Instituto Superior Técnico para a Investigação e Desenvolvimento, University of Lisbon, Av. António José de Almeida, n.º 12, 1000-043 Lisbon, Portugal
e-mail: mpaulamendes@tecnico.ulisboa.pt

I. Flores Colen
CERIS, Instituto Superior Técnico, University of Lisbon, Av, Rovisco Pais 3, 1049-001 Lisbon, Portugal

M. Torres González and C. Rubio Bellido (eds.), *Towards Low and Positive Energy Buildings*, Green Energy and Technology, https://doi.org/10.1007/978-3-031-70851-0_1

adaptation and increasing sustainability. To provide a clearer understanding of these principles, the text also presents several illustrative examples.

1 Climate Change and Extreme Weather Events in Europe

July 2023 marked a historic moment, as the global average temperature surpassed all previous records for any month [1]. Anthropogenic greenhouse gas (GHG) emissions are at an all-time high, and even if they were halted, the climate impacts that have already been initiated would persist for decades [2]. The increase in GHGs in the atmosphere increases humidity and intensifies warming [3], creating a feedback loop that leads to increased temperatures and higher absorption of water vapor by the atmosphere. This exacerbates climate change, resulting in short-term and long-term variations [2].

Since around 1950, various extreme weather and climate events have been documented, encompassing heat waves, droughts, floods, cyclones, and wildfires [4]. Consequently, since 2017, extreme weather has consistently ranked as the most likely threat to the economy in the Global Risks Report [5]. Extreme weather events can significantly impact society more than the gradual long-term increase in average temperature [6, 7]. There is a strong certainty that their frequency and magnitude will increase due to climate change [4]. Europe, in particular, has witnessed more rapid warming than any other continent in recent decades [8], resulting in unprecedented heat waves over the past twenty years [9].

Moreover, over the last two decades (2001–2020) in Europe, there were 951 weather-related disaster events, encompassing floods (41%), extreme temperatures (23%), storms (27%), and wildfires (6%). Among these events, heat waves accounted for the highest number of recorded deaths (146,000), followed by cold waves (3,800) and floods (2,142) [10].

Cold waves can have profound impacts, leading to unusual snow accumulations and posing significant risks such as fatalities caused by exceptionally cold weather and avalanches. These events directly affect various sectors, including energy demand, transport systems, and social emergency protection [11].

The severe impact of cold waves was starkly evident during the winter of 2005–2006, marked by Ukraine's report of 738 deaths within 3 weeks and Poland's recording of 150 cold-related deaths. Moreover, this period saw over 80 avalanche-related fatalities, with the French Alps experiencing the highest number ever recorded [12, 13]. The winter of 2009/2010 witnessed severe cold spells in Northwestern Europe, while milder conditions prevailed in Southern Europe [11]. The extreme cold weather caused disruptions in Eurostar train services, other railways, airports, and roads [14]. Other European countries, such as Hungary, where 48 fatalities were recorded; the Russian Federation, with 116 deaths; and Romania, with 48 fatalities, also reported additional casualties due to the severe winter conditions [15]. Recently, at the beginning of 2024, Finland and Sweden recorded their coldest winter temperatures on January 2, reaching minus 40 °C, as a cold wave swept across the Nordic region. Two fatalities were reported due to an avalanche in the Finnish Lapland region. This extreme event disrupted transportation, resulting in the closure of a major highway in southern Norway and the suspension of ferry operations. In Sweden, thousands were left without power, and many endured hours stuck in their cars along congested highways, facing the extremely cold temperatures [16].

However, across Europe, all regions are observing rising mean temperatures and decreasing frequency of cold spells [17]. Heat waves and heat-related events can profoundly affect people, leading to increased risks of heatstroke, cardiovascular and cerebrovascular diseases, respiratory illnesses, and premature mortality [18]. Currently, vulnerable groups—adults over 65 and children under one year old—experience double the number of heat wave days compared to the period from 1986 to 2005. Additionally, the combined impact of heat waves and droughts in 2021 led to an extra 127 million people worldwide facing moderate to severe food insecurity compared to the period from 1981 to 2010 [19].

Between 2000 and 2022, the western and southern regions of Europe experienced more frequent heat waves, with rates of 39% and 36%, respectively [15], resulting in several causalities. The summer heat waves of 2003 in western and central Europe resulted in an estimated 34,177 deaths, while those in Eastern Europe and Russia in 2010 caused 11,000 casualties [20–23]. In 2022, high-temperature records were notably broken, particularly in Spain, Italy, and France. June 2022 marked the third warmest June globally on record [24]. During that summer, a prolonged series of intense heat waves caused extreme temperatures, drought, and increased fire activity, resulting in almost 62,000 fatalities across Europe, including 2,212 in Portugal, 11,324 in Spain, and 18,010 in Italy [25]. The increased frequency of heat waves has caused an increase in energy demands as people rely more on heating, ventilation, and air-conditioning (HVAC) systems and other cooling systems. This increase in energy usage strains the electric grid, potentially leading to power outages [26]. However, when evaluating total economic losses in Europe, extreme temperatures accounted for just 9%, while floods and storms constituted larger portions at 50% and 29%, respectively [10].

Europe has witnessed several major floods, significantly damaging infrastructure, property, and loss of life. According to the EM-DAT database [15], between 2012 and 2022, Europe faced 56 riverine flood-related disasters and 40 flash flood incidents. Eastern Europe accounted for the majority (50%) of riverine floods, with Russia experiencing the highest frequency (25%), followed by Bulgaria (21%). In Southern Europe, flash floods constituted around 65% of incidents, with Italy reporting the highest frequency (31%).

In July 2021, torrential rainfall triggered extreme flash floods across Germany, Belgium, Luxembourg, and the Netherlands. These floods resulted in over 200 fatalities (more than 170 in Germany and 32 in Belgium), extensive infrastructure damage (including water supply, discharge lines, and the power grid), incurring losses exceeding 54 billion US Dollar, and displacing thousands of people. These catastrophic events have underscored the inadequate preparation in many parts of Western Europe to handle heavy rainfall [24, 27].

Numerous other flood events occurred in southern Europe during 2023, resulting in major economic losses and some loss of life. In May, two intense rain events two weeks apart, on 1–3 May and 16–18 May, affected the Emilia-Romagna region of northern Italy. Two-day totals exceeded 250 mm in the second event and a record total of 260.8 mm in 48 h at Monte Albano.

In November 2019, in Venice, Italy, apart from a storm surge, three other significant factors contributed to a flood: a high tide (for instance, one of the tidal peaks: + 0.26 m), an exceptionally high level of the Mediterranean Sea, and a deep, small-scale atmospheric pressure minimum that developed over the Adriatic. The storm lasted less than an hour but caused extensive damage to the town [28].

Floods and storms are among the most economically costly natural hazards that can be influenced by climate change [29]. Extratropical cyclones, marked by robust winds, precipitation, temperature shifts, and cloud cover, are predominant in the mid-latitudes [30]. Storms across Europe largely stem from these cyclones, originating when warm subtropical air interacts with polar air over the Atlantic Ocean [31]. Between 2012 and 2022, Europe faced 35 disasters associated with these cyclones. While Southern Europe was affected by seven storms, the majority occurred in Western Europe (40%), with France reporting the highest frequency at 79% [15]. In 2009, an extratropical storm named "Klaus" had significant effects in Spain, France, and Italy, resulting in 14, 11, and 3 deaths, respectively, resulting in 22.9 billion US Dollars in damages [15]· In early September 2023, a major cyclone, "Storm Daniel", struck the eastern Mediterranean bringing exceptional rainfall. The Thessaly region of Greece recorded the highest rainfall amounts. Southern Bulgaria and Turkey also experienced extreme rainfalls and flooding, while simultaneously, a second system in the west caused significant flash flooding in Spain. Zagora Pelion in Greece received 759.6 mm of rain on September 5, with a 5-day total of 1096.2 mm from September 4–8. In Bulgaria, Kos received 329 mm in 16 h on September 4 and 5. The flooding led to 19 reported deaths in Greece and Bulgaria, with major disruptions to transportation and severe agricultural losses, including extensive silt coverage on highly productive land [32].

Convective storms are characterized by the presence of thunder, lightning, heavy rain, hail, strong winds, and sudden temperature changes [28]. The rainfall from a convective storm is notably intense and localized. These storms caused 55 reported disasters, with Western Europe experiencing the highest frequency at 40%, and severe storms accounting for 27.5% of the occurrences [15]. In October 2020, convective storm Alex brought heavy rainfall and strong winds, leading to flash floods and landslides in southeastern France and northwestern Italy. The storm inflicted substantial damage to infrastructure (including, bridges, roads, electric power lines, landlines, and some mobile phone services) and buildings in these areas, resulting in loss of life [27, 31].

These extreme weather events are exacerbated by climate change, which is now recognized as the primary twenty-first-century global health threat [10, 11]. Rising temperatures are also likely to increase the transmission potential of many vector-borne, waterborne, foodborne, and airborne infectious diseases, exposing larger populations to potentially fatal infectious diseases such as dengue, Zika virus, malaria, chikungunya, West Nile virus, and non-cholera Vibrio pathogens [19]. The West Nile virus (WNV), a climate-sensitive disease transmitted by mosquitoes, primarily circulates among spring migration birds from infected regions of Sub-Saharan Africa and can be transmitted to humans through *Culex* mosquitoes. Since 2010, WNV has been reported in 14 EU countries, including Austria, Bulgaria, Croatia, Cyprus, Czechia, France, Greece, Hungary, Italy, the Netherlands, Portugal, Romania, Slovenia, and Spain. Some factors contributing to WNV outbreaks include rising winter temperatures, warmer summer temperatures, and shrinking water resources during some seasons [33].

The economic toll of climate change is significant and includes loss of life, property damage, and ruptures in transportation and energy supply. Climate change impacts reach beyond extreme weather events and rising sea levels. It affects the emergence of infectious diseases like WNV, which can thrive in the changing environmental conditions caused by climate change. It is essential to consider the interconnected nature of climate change impacts when developing policies and strategies to address this pressing issue.

2 Potential Impact of Climate Change and Needs of Adaptation

The Intergovernmental Panel on Climate Change (IPCC) defines adaptation as an adjustment to climate change. This applies to human systems, where human intervention may help adapt to anticipated climate scenarios, mitigating harm or taking advantage of opportunities. On the other hand, vulnerability encompasses various concepts and elements, such as sensitivity or susceptibility to harm, and the inability to cope and adapt to changes.

While Europe exhibits a relatively high adaptive capacity compared to other global regions, adaptation is constrained by various physical, social, economic, and technological factors [17]. The European Union population is distributed among large cities (40%), towns and suburban areas (35%), and rural regions (25%) [34]. In Europe, the vulnerability of the building stock and critical infrastructures to climate change is influenced by urban layout, socioeconomic activities, economic conditions, and governance factors [17]. This vulnerability is further underscored by the fact that nearly 50 million Europeans reside within 10 m above mean sea level, placing a significant population at risk as sea levels continue to rise [35], with soft cliffs and beaches being the most affected by erosion [17]. Moreover, unplanned urban development driven by tourism significantly shapes present-day flood risks in coastal regions [36] and near rivers [37]. These flood-prone areas are particularly attractive to both inhabitants and socioeconomic activities, exacerbating the challenges posed by climate change [37]. This exposition can damage critical infrastructure, housing, economic losses, and the displacement of coastal communities.

Many coastal regions of the North Sea, especially in the southern areas, are particularly vulnerable to rising sea levels due to the region's strong tidal regime and the impact of storm surges. Numerous coastal populations risk flooding, especially when high tides coincide with storm surges and/or elevated river flows [17, 35, 38]. Furthermore, coastal floods alone could exceed 1 trillion euros per year [39].

Extreme heat is increasingly common, particularly in southern and Western Europe, putting a significant portion of the population at risk of heat stress [39]. Elevated urban temperatures exacerbate infrastructure overheating and failures, potentially intensifying the concentration of harmful air pollutants like ozone [40]. Additionally, in southern Europe, the risks associated with wildfires are already categorized as "critical", necessitating an urgent call to action [39].

Table 1 describes the probable climate hazards, and Table 2 summarizes the potential effects of extreme weather events, and climate change on the built environment in Europe. The risks of damage to building facades vary across different locations and depend on a combination of climate drivers (e.g., higher temperature combined with higher precipitation) and building design and construction parameters. These include material types and construction solutions, such as wall insulation, wall thickness, and building height, among others [41, 42].

This model of urban development characterized by construction in floodplains, extensive surface sealing, and limited green spaces, as well as urban sprawl encroaching on areas prone to wildfires or landslides, exacerbates present and projected impacts of climate change [43]. To effectively address these pressing challenges, imperative strategies must be implemented. These include the enhancement of building design and adherence to regulations and strategies, such as:

- Effective implementation of the Floods Directive 2007/60/EC [44], which is tailored to assess and manage flood risk.
- Utilization of European and international standards and technical specifications pertinent to security and resilience measures applicable to building environment adaptation to climate change.
- Integration of EU climate and energy policies to more comprehensively incorporate climate adaptation into the planning and execution of measures within the energy sector [39].
- Implementing alternative cooling methods to conserve energy during heat waves and employing rain harvesting and storage systems to alleviate the impact of drought events are crucial measures [45]. Buildings with no active cooling systems, intermittent access to power, or frequent power outages are prone to overheating issues [46].
- Reintegrate nature into urban environments; it is important to use nature-based solutions (NBSs), since they can be designed to augment biodiversity while simultaneously mitigating the adverse impacts of climate change [47, 48].
- Minimizing the environmental footprint of buildings involves adopting a life-cycle perspective and embracing circularity principles. This entails optimizing resource use and promoting sustainability throughout the construction sector [49].

Table 1 Relevance of climate variables for climate change adaptation in Northern, Southern, and Central Europe

(a)

Climate hazard	North of Europe		Central Europe		South Europe	
	Observed trend	Projected trend	Observed trend	Projected trend	Observed trend	Projected trend
Winter rainfall [50–52]	Significant positive trend over 1902–2020	Increasing (medium confidence)	Trends are less clear in continental Europe			Decreasing (medium confidence)
Heavy precipitation events [51–55]	Increase in winter since the 1950s. Increased snow load is observed across all seasons, especially winter. There are clear indications of increased snow load in northern Sweden, Finland, and Norway. Scandinavia is experiencing increases in heavy precipitation during summer		Increase since the 1950s in some parts of west-central Europe and Russia, especially in winter (medium confidence)	Increase in heavy precipitation over winter	Inconsistent trends, in particular in summer	Increase in heavy precipitation in winter
Warming in winter [51, 56]	High warming has been observed over Scandinavia	Warming stronger in winter	Warming stronger in winter			The Iberian Peninsula (IBP) warmed mostly in summer

(continued)

Table 1 (continued)

(a)

Climate hazard	North of Europe		Central Europe		South Europe	
	Observed trend	Projected trend	Observed trend	Projected trend	Observed trend	Projected trend
Droughts [50, 51, 53, 55, 57, 58]	Southernmost regions of Sweden, a slight drying trend in spring and summer was found			Streamflow drought and water scarcity are more severe and persistent (medium confidence)	Evidence of increased severity of climatic droughts and decreased surface water resources in the IBP	Streamflow drought and water scarcity are more severe and persistent (medium confidence). Further exacerbation of groundwater depletion and groundwater level reduction

(continued)

Table 1 (continued)

(b)

Climate hazard	North of Europe		Central Europe		South Europe	
	Observed trend	Projected trend	Observed trend	Projected trend	Observed trend	Projected trend
Extreme heat days [51, 52, 59, 60]		Very likely increase in hot days (decrease in cool days) Increase of about 2–4 °C	Very likely increase of observed trends in the magnitude and frequency of high maximum temperatures has been observed consistently	Very likely increase of observed trends in the magnitude and frequency of high maximum temperatures has been observed consistently	Likely strongest and most hot days' significant trends in Iberian peninsula and southern France	Increase of about 4–6 °C

(continued)

Table 1 (continued)

(b)

Climate hazard	North of Europe		Central Europe		South Europe	
	Observed trend	Projected trend	Observed trend	Projected trend	Observed trend	Projected trend
Extreme wind [51, 61]		Increase in occurrence of extreme winds as we go further in the future (RCP8.5 forcing)		Increase in the occurrence of extreme winds as we go further in the future (RCP8.5 forcing)		Increase in the occurrence of extreme winds as we go further in the future (RCP8.5 forcing)
Extreme wind and rain extratropical cyclones [62]	High percentages of co-occurring wind and precipitation extremes along the west coast of Norway		High percentages of co-occurring wind and precipitation extremes along northwestern central Europe		High percentages of co-occurring wind and precipitation extremes along the west coast of the IBP	

Table 2 Potential impact of climate hazards on the built environment

Climate hazard	Potential impact on the built environment
Winter rainfall [41, 50, 57, 63]	Deeper penetration of water into the building fabric Moisture in buildings can accelerate material decay and mold growth, leading to significant economic and health issues Masonry will stay saturated longer, increasing physical and chemical stress on stonework, leading to more decay and vulnerability to frost damage
Heavy Precipitation Events [37, 40, 51]	Increased risk of water damage due to inadequate rainwater system capacity Higher snow loads could jeopardize building safety in some northern or mountainous regions Floods shut down hydroelectric projects, resulting in greater use of bunker fuel, fuel oil, and natural gas
Warming in winter [40, 64–66]	Decay or biological degradation of wooden structures in construction. The extent of the seasonal activity of pests and plant diseases. Accelerate the degradation of painted finishes to timber elements such as sash and case windows and doors
Droughts [67]	It increases the likelihood of forest fires and the significant risk of buildings burning down or becoming uninhabitable In regions with buildings on swelling clay, drought can cause the ground to shrink, leading to subsidence and damage Buildings in areas with groundwater lowering and soil settlement may face foundation problems, roads, and bridges Historic wall finishes may deteriorate, leading to discoloration and damage (i.e., fading, cracking, weakening, and disintegration)
Extreme heat days [68–73]	Reduction of the longevity of materials (e.g., wood, plastic, stone, renders, and paint) which are degraded by solar radiation and high temperatures Wildfires - buildings burned down or rendered inhabitable Buildings designed for certain thermal conditions will require modifications to cope with warmer climates Damage to road and rail transport systems
Extreme wind [61, 74, 75]	Accurate evaluation of extreme winds in coastal and mountainous regions is critical to safely design large buildings and infrastructure. Extreme winds significantly affect structures, especially bridges and tall buildings
Extreme wind and rain extratropical cyclones [62, 75–77]	Rainwater penetration can affect the building envelope's thermal performance, facade deterioration, and occupants' health, including respiratory problems like asthma. Current international standards for facades may not effectively address wind-driven rain (WDR), leading to surface erosion and moisture issues in historic buildings

3 Circularity and Energy Efficiency and Buildings Toward Climate Change Adaptation

Growing concern about sustainability and the environmental impact of construction has driven the adoption of more conscious and efficient practices in the construction sector. The life cycle of buildings is associated with significant environmental and socioeconomic impacts. However, adopting circularity and energy efficiency principles can significantly reduce these impacts. The full life cycle of buildings in the EU accounts for approximately 50% of the total energy use, 40% of the total GHG emissions, 50% of the raw material extraction, and a third of all water use. The linear economic model is inadequate for achieving sustainability and cleaner resource production [78]. Circularity, a principle derived from the circular economy, addresses ways to minimize resource waste, reduce GHG emissions, and promote a smarter, more sustainable use of natural resources [79]. Energy efficiency is also crucial, focusing on reducing energy consumption in buildings without compromising comfort and functionality.

The European Union recognizes that achieving climate neutrality will require extensive building stock renovation. This involves reducing buildings' energy demands, decarbonizing the energy used, and mitigating emissions from a life-cycle perspective [80]. For instance, when renovating, design-for-disassembly (DfD) products can disrupt the conventional linear paradigm of "take-use-dispose" by simplifying the reuse and recycling of building components at the end of their useful life. Consequently, components and parts can be seamlessly reused, recycled, recovered for energy, or diverted from the waste stream [81]. Other renovation actions can be undertaken to increase building lifespans, reduce material consumption, and integrate new-generation materials. These actions promote the adoption of innovative, high-performance, durable, sustainable, and resilient materials [82, 83]. Figure 1 presents renovation actions, highlighting that extending building lifespans has the greatest potential for material and GHG cumulative savings. This extension translates into a lower demand for new construction, which is highly material and GHG intensive. This is followed by reduced material consumption, mainly achieved through boosting the high-quality recycling of materials. Although most long-lasting materials are more GHG-intensive during production than traditional materials, they require less frequent renovation, thereby avoiding using materials and their embedded GHG emissions in the long term. Lastly, although the impact is smaller, using bio-based materials and green solutions still offers significant benefits. The reasons for having less potential than increasing lifespans and reducing material consumption are that bio-based materials typically need to be replaced more frequently and are more GHG-intensive to produce per unit than existing alternatives. However, the cost of GHG emissions is low, and they can offer a wide range of benefits if specific considerations like maintenance, adaptability, and soil and water resources are properly addressed [48]. These benefits include recreational opportunities, climate adaptation to heat stress, air temperature cooling, carbon sequestration, storm water treatment and management, and increased biodiversity [84].

The benefits of retrofitting existing buildings in terms of energy efficiency, especially through passive solutions, and material conservation, are clear [85, 86]. These benefits include improved indoor comfort [87], enhanced hygrothermal performance [41], and a decreased need for new construction, which is more detrimental in terms of material usage, energy consumption, soil sealing, habitat disturbance, and emission of pollutants [88].

Buildings must be adapted, designed, and constructed to address the effects of global warming, including increased heat island effect [48], rising cooling demands,

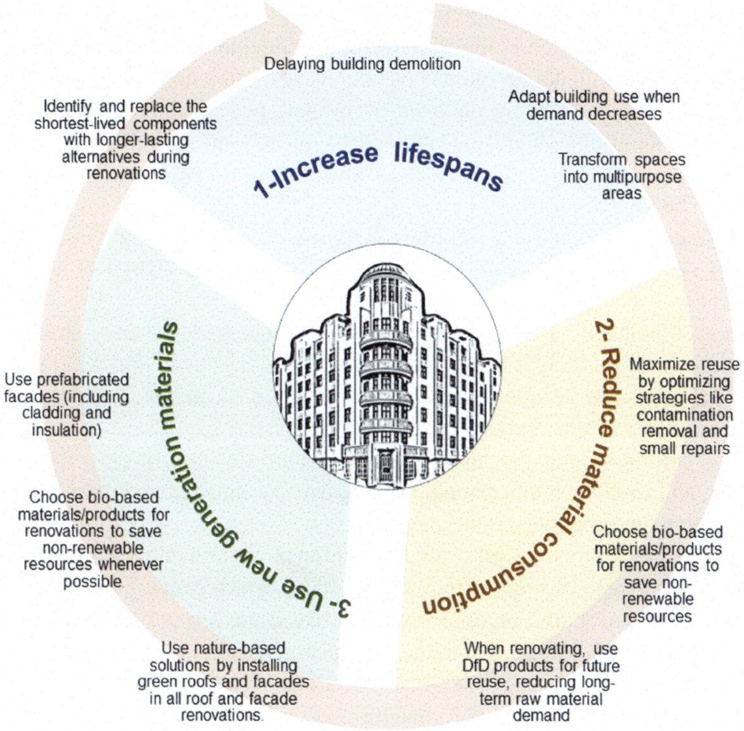

Fig. 1 Ten circular renovation actions for increasing 1—building lifespans, 2—reducing material consumption, and 3—incorporating new-generation materials

and water scarcity. Moreover, they should provide safety and resilience against adverse climatic events.

As the impact of climate change intensifies, considering energy efficiency and circularity in building design, operation, and renovation is vital for ensuring sustainability and resilience. Here are some key strategies to improve the buildings' resilience to climate change [48, 89–93].

- Encourage renovation of existing buildings to improve energy efficiency and adaptability to extreme weather events, such as winter rainfall, heavy precipitation, winter warming, droughts, extreme heat, extreme winds, and storms, across northern, central, and southern Europe.
- Address the specific challenges of heat and sun exposure (e.g., in Mediterranean countries) by focusing on shading, ventilation, and cooling strategies and the challenges posed by cold climate and high energy demands (e.g., in Scandinavian countries) by considering suitable insulation, heating systems, and materials for cold weather conditions. Integrate strategies that enhance the building's resilience to climate change impacts and potential hazards, considering historical aesthetic requirements, material durability, and DfD products.

- Choose thermal insulation materials based on the microclimate conditions of the building's location. Factors like air temperature fluctuations, humidity levels, and insulation requirements should be considered. Although most organic fossil fuel-derived materials are being replaced by organic plant—or mineral-based ones, some considerations should be outlined to understand if their implementation could be potentially beneficial or disadvantageous in mitigating and adapting to climate change scenarios.
- Incorporate multifunctional building features, such as daylight optimization, natural ventilation, and rainwater harvesting, to enhance energy efficiency and resident well-being.
- Promote a balance between passive and active energy efficiency strategies by choosing energy-efficient lighting systems, HVAC systems, and windows. Integrate nature-based solutions and microgeneration. Aim for the highest green building certifications during the design stage.
- Update building standards and codes to include design criteria responsive to heating and cooling requirements and passive cooling features that can reduce energy consumption and enhance comfort.
- Establish long-term measures for facilitating a sustainable energy transition. These encompass various initiatives, such as offering energy-efficient mortgages to support building renovations, advocating for public investments in energy-efficient buildings, and addressing perceived investment risks.
- Engage local communities and involve citizen science participation to raise awareness about climate-related changes in urban areas. Educate stakeholders on circular economy principles and use building energy modeling (e.g., BEM) with comprehensive data input to address climate change impacts. Leverage digital tools to inform adaptive urban planning considering climate resilience, hazard mitigation, and circular economy strategies.
- Foster a collaborative approach between stakeholders (e.g., local communities, city officials, urban planners, and researchers) and/or support regulatory frameworks. This alignment ensures that policies and strategies align with the overarching goal of sustainable development.
- Implement wireless sensor networks for real-time data collection from multiple points in the building or prototype systems (in innovative solutions), enabling continuous monitoring and KPIs evaluation over time (e.g., energy consumption reduction, energy use intensity, thermal comfort improvement, and renewable energy growth rate). Leverage digital tools to inform adaptive urban planning, considering climate resilience, hazard mitigation, and circular economy strategies. Transmit data to cloud-based platforms for real-time analysis and backup. This enhances data accessibility and facilitates timely decision-making based on real-time information.

Most buildings across Europe have an average lifespan of over 50 years and were constructed before thermal standards and energy efficiency regulations were implemented. Investing in building refurbishment is crucial for reducing energy consumption and GHG emissions in the EU. In Portugal, approximately 67% of the building stock predates 1990, before the thermal regulation, which mandated thermal insulation requirements be included in the design of new buildings [94]. Consequently, a significant part of the Portuguese buildings is deemed to have unsatisfactory thermal performance, contributing significantly to energy poverty and indoor discomfort among residents [94]. Investing in building renovation, especially for older buildings with poor thermal performance, can significantly reduce energy consumption and improve indoor comfort and residents' well-being. Enhanced hygrothermal performance also decreases the degradation risk of facades. Besides wall orientation, other factors or their combinations can significantly influence the hygrothermal response of building envelopes, such as the presence of insulation, the materials, and thickness of the walls [41, 42].

The growing concern about the environmental impact of construction has urged the adoption of more conscious and efficient practices within the construction sector. With buildings accounting for a significant portion of energy use, GHG emissions, raw material extraction, and water consumption in the EU, addressing these impacts is important. Circularity presents a promising approach to reducing waste and GHG emissions, promoting the sustainable use of resources, particularly within the highly material-intensive and energy-demanding building sector. Energy efficiency is central due to the reciprocal relationship between buildings and climate change. Integrating energy efficiency and circularity principles into building design and operation is vital for resilience against climate change and climate mitigation.

4 Conclusions

The full life cycle of buildings in the EU accounts for significant environmental and socioeconomic impacts, including high-energy use, GHG emissions, raw material extraction, and water use. The traditional linear economic model is inadequate for achieving the 2050 climate goals.

This chapter outlines the potential effects of climate change and extreme weather events on Europe's built environment. It examines various factors, including winter rainfall, heavy precipitation, winter warming, droughts, extreme heat, extreme wind, and extratropical cyclones across northern, central, and southern Europe. In addition, it identifies observed and projected trends and explains how these climate factors can affect the built environment in these regions. Climate change also influences the spread of infectious diseases such as WNV.

Implementing circularity in the building sector, which is highly material-intensive, involves employing various strategies and practices. Central to this approach is the recognition that actions aimed at increasing building lifespans are paramount. The European Union acknowledges that achieving climate neutrality requires extensive renovation of the building stock, which encompasses reducing energy demands, decarbonizing energy use, and mitigating emissions from a life-cycle perspective. This involves increasing the lifespans of buildings, utilizing new generations of materials, and reducing material consumption. Through the promotion of innovative, high-performance, durable, sustainable, and resilient materials, these actions drive forward the transition toward a more sustainable built environment.

Energy efficiency emerges as crucial for reducing the carbon footprint, as it focuses on lowering energy consumption in buildings without compromising comfort and functionality. This chapter underscores the significant benefits associated with retrofitting existing buildings, particularly in terms of enhancing energy efficiency and conserving materials. Retrofitting improves indoor comfort and reduces the need for new construction, which is inherently more resource-intensive in terms of material usage, energy consumption, and pollutant emissions. Furthermore, it underscores the imperative of adapting, designing, and constructing buildings to mitigate the effects of global warming, which include addressing challenges such as the urban heat island effect, extreme wind and rain events, escalating cooling demands, and water scarcity. Additionally, buildings must be designed to be adaptable to new uses and incorporate materials/products designed for disassembly (DfD). As the impacts of climate change escalate, integrating energy efficiency and circularity into building design, operation, maintenance, and renovation processes becomes increasingly imperative.

Acknowledgements M. P. Mendes acknowledges the Portuguese Foundation for Science and Technology (FCT) for strategic funding and research unit CERENA (UIDB/04028/2020).

I. Flores Colen acknowledges the FCT and research unit CERIS (UIDB/04625/2020).

References

1. Copernicus Climate Change Service (C3S) (2023) The warmest month in Earth's recent history. Monthly Climate Bulletin July 2023, pp 3–9
2. UNFCCC (2022) Climate-related risks and extreme events
3. Hansen K (2008) Water vapor confirmed as major player in climate change. NASA—ScienceDaily
4. IPCC (2014) Contribution of working groups I, II and III to the fifth assessment report of the inter-governmental panel on climate change. In: Core writing team RKP and LAM (eds) climate change 2014: synthesis report. Geneva, Switzerland
5. World Economic Forum (2021) The global risks report
6. Walsh JE, Ballinger TJ, Euskirchen ES et al (2020) Extreme weather and climate events in northern areas: a review. Earth Sci Rev 209:103324. https://doi.org/10.1016/j.earscirev.2020.103324
7. Mendes MP, Rodriguez-Galiano V, Aragones D (2022) Evaluating the BFAST method to detect and characterise changing trends in water time series: A case study on the impact of droughts on the Mediterranean climate. Sci Total Environ 846:157428. https://doi.org/10.1016/j.scitotenv.2022.157428
8. C3S (2022) Climate indicators temperature. In: Copernicus climate change service
9. Becker FN, Fink AH, Bissolli P, Pinto JG (2022) Towards a more comprehensive assessment of the intensity of historical European heat waves (1979–2019). Atmos Sci Lett 23:e1120. https://doi.org/10.1002/asl.1120
10. Centre for Research on the Epidemiology of Disasters (2021) CRED Crunch 64—Extreme weather events in Europe. 2
11. Cattiaux J, Vautard R, Cassou C et al (2010) Winter 2010 in Europe: a cold extreme in a warming climate. Geophys Res Lett 37:1–6. https://doi.org/10.1029/2010GL044613

12. Graham RJ, Gordon C, Huddleston MR et al (2006) The 2005/06 winter in Europe and the United Kingdom: part 1—how the met office forecast was produced and communicated. Weather 61:327–336. https://doi.org/10.1256/wea.181.06
13. Scaife AA, Knight JR (2008) Ensemble simulations of the cold European winter of 2005–2006. Q J R Meteorol Soc 134:1647–1659. https://doi.org/10.1002/qj.312
14. BBC UK (2009) Thousands freed from channel tunnel after trains fail. In: News
15. Centre for Research on the Epidemiology of Disasters (CRED) (2023) Université catholique de Louvain (UCLouvain) Emergency Events Database (EM-DAT)
16. Bryant M (2024) Finland avalanche leaves two dead as extreme cold grips Nordic countries. The Guardian
17. Bednar-Friedl B, Biesbroek R, Schmidt DN et al (2022) Europe. In: Pörtner HO, Roberts DC, Tignor M et al (eds) Climate change 2022: impacts, adaptation and vulnerability. Contribution of working group II to the Sixth assessment report of the intergovernmental panel on climate change. Cambridge University Press, Cambridge, UK and New York, NY, USA, pp 1817–1927
18. Naumann G, Russo S, Formetta G et al (2020) Global warming and human impacts of heat and cold extremes in the EU. Publ Office Eur Union. https://doi.org/10.2760/47878
19. Romanello M, di Napoli C, Green C et al (2023) The 2023 report of the <em>Lancet</em> Countdown on health and climate change: the imperative for a health-centred response in a world facing irreversible harms. Lancet 402:2346–2394. https://doi.org/10.1016/S0140-6736(23)01859-7
20. UNEP (2004) Environment alert bulletin: impacts of summer 2003 heat wave in Europe. 3
21. Poumadère M, Mays C, Le Mer S, Blong R (2005) The 2003 heat wave in France: Dangerous climate change here and now. Risk Anal 25:1483–1494. https://doi.org/10.1111/j.1539-6924.2005.00694.x
22. Garcia-Herrera R, Díaz J, Trigo RM et al (2010) A review of the european summer heat wave of 2003. Crit Rev Environ Sci Technol 40:267–306. https://doi.org/10.1080/10643380802238137
23. Shaposhnikov D, Revich B, Bellander T et al (2014) Mortality related to air pollution with the moscow heat wave and wildfire of 2010. Epidemiology 25:359–364. https://doi.org/10.1097/EDE.0000000000000090
24. The Lancet Regional Health—Europe (2021) Germany floods—a warning for future extreme weather events. The Lancet Regional Health Europe 7:100194. https://doi.org/10.1016/j.lanepe.2021.100194
25. Ballester J, Quijal-Zamorano M, Méndez Turrubiates RF et al (2023) Heat-related mortality in Europe during the summer of 2022. Nat Med 29:1857–1866. https://doi.org/10.1038/s41591-023-02419-z
26. Stone B, Mallen E, Rajput M et al (2021) Climate change and infrastructure risk: Indoor heat exposure during a concurrent heat wave and blackout event in Phoenix Arizona. Urban Climate 36:100787. https://doi.org/10.1016/j.uclim.2021.100787
27. Dolejš M, Raška P, Kohnová S et al (2022) On the right track of flood planning policy? Land uptake in Central-European floodplains (1990–2018). Landscape Urban Plan 228:104560. https://doi.org/10.1016/j.landurbplan.2022.104560
28. RMetS (2020) What are convective storms? In: Royal meteorological society
29. European Commission (2020) Overview of natural and man-made disaster risks the European Union may face
30. Pinto JG, Ludwig P (2020) Extratropical cyclones over the North Atlantic and Western Europe during the last glacial maximum and implications for proxy interpretation. Climate Past 16:611–626. https://doi.org/10.5194/cp-16-611-2020
31. European Commission (2014) Commission staff working document: overview of natural and man-made disaster risks in the EU. Brussels, 8.4.2014. Final, SWD 134
32. World Meteorological Organization (2023) Significant weather & climate events
33. Watts MJ, Monteys VS, Mortyn PG, Kotsila P (2021) The rise of West Nile Virus in Southern and Southeastern Europe: a spatial–temporal analysis investigating the combined effects of climate, land use and economic changes. One Health. 1(13):100315. https://doi.org/10.1016/j.onehlt.2021.100315

34. Commission E, Centre JR, Jacobs-Crisioni C et al (2018) European territorial trends: facts and prospects for cities and regions. Publications office
35. McGranahan G, Balk D, Anderson B (2007) The rising tide: assessing the risks of climate change and human settlements in low elevation coastal zones. Environ Urban 19:17–37. https://doi.org/10.1177/0956247807076960
36. Fraser A, Pelling M, Scolobig A, Mavrogenis S (2020) Relating root causes to local risk conditions: a comparative study of the institutional pathways to small-scale disasters in three urban flood contexts. Glob Environ Change 63:102102. https://doi.org/10.1016/j.gloenvcha.2020.102102
37. Postacchini M, Zitti G, Giordano E et al (2019) Flood impact on masonry buildings: the effect of flow characteristics and incidence angle. J Fluids Struct 88:48–70. https://doi.org/10.1016/j.jfluidstructs.2019.04.004
38. Vousdoukas MI, Mentaschi L, Voukouvalas E et al (2018) Climatic and socioeconomic controls of future coastal flood risk in Europe. Nat Clim Chang 8:776–780. https://doi.org/10.1038/s41558-018-0260-4
39. European Environment Agency (2024) European climate risk assessment executive summary EEA report 01/2024. Luxembourg
40. Dodman D, Hayward B, Pelling M et al (2022) Cities, settlements and key infrastructure. In: Climate change 2022: impacts, adaptation and vulnerability. contribution of working group II to the sixth assessment report of the intergovernmental panel on climate change. Cambridge University Press, Cambridge, UK and New York, NY, USA
41. Vandemeulebroucke I, Kotova L, Caluwaerts S, Van Den Bossche N (2023) Degradation of brick masonry walls in Europe and the Mediterranean: advantages of a response-based analysis to study climate change. Build Environ 230:109963. https://doi.org/10.1016/j.buildenv.2022.109963
42. Defo M, Lacasse MA (2021) Effects of climate change on the moisture performance of tallwood building envelope. Buildings. https://doi.org/10.3390/buildings11020035
43. EEA (2020) Urban adaptation in Europe: how cities and towns respond to climate change
44. European Parliament and the Council of the European Union (2007) Directive 2007/60/EC of the European parliament and of the council of 23 Oct 2007 on the assessment and management of flood risks. Off J Eur Union 288:27–34
45. Mendes MP, Flores-Colen I, Bertolin C (2021) Discussion of performance criteria for thermal insulating solutions for buildings facades under different climate change projections. In: Itecons—Instituto de Investigação e Desenvolvimento Tecnológico para a Construção, Energia A e S (eds) CEES 2021—International conference on construction, Energy, Environ Sustain. Coimbra, Portugal
46. Laouadi A, Bartko M, Lacasse MA (2020) A new methodology of evaluation of overheating in buildings. Energy Build 226:110360. https://doi.org/10.1016/j.enbuild.2020.110360
47. European Commission (2020) EU biodiversity strategy for 2030 - Bringing nature back into our lives COM/2020/380 final
48. Pereira C, Flores-Colen I, Mendes MP (2024) Guidelines to reduce the effects of urban heat in a changing climate: green infrastructures and design measures. Sustain Dev 32:57–83. https://doi.org/10.1002/sd.2646
49. European Commission (2020) A renovation wave for Europe—greening our buildings, creating jobs, improving lives COM 662 final
50. Ionita M, Nagavciuc V, Scholz P, Dima M (2022) Long-term drought intensification over Europe driven by the weakening trend of the Atlantic meridional overturning circulation. J Hydrol Reg Stud 42:101176. https://doi.org/10.1016/j.ejrh.2022.101176
51. Hewitson B, Janetos AC, Carter TR et al (2014) Regional Context. In: Barros VR, Field CB, Dokken DJ et al (eds) Climate change 2014: impacts, adaptation, and vulnerability. Cambridge University Press, Cambridge, United Kingdom and New York, NY, pp 1133–1197
52. Nikulin G, Kjellstro ME, Hansson U et al (2011) Evaluation and future projections of temperature, precipitation and wind extremes over Europe in an ensemble of regional climate simulations. Tellus A Dyn Meteorol Oceanogr 63:41–55. https://doi.org/10.1111/j.1600-0870.2010.00466.x

53. Larsson Ivanov O, Bärring L, Wilcke RAI (2022) Climate change impact on snow loads in northern Europe. Struct Saf 97:102231. https://doi.org/10.1016/j.strusafe.2022.102231
54. Croce P, Formichi P, Landi F et al (2018) The snow load in Europe and the climate change. Climate Risk Manag 20:138–154. https://doi.org/10.1016/j.crm.2018.03.001
55. Schipper ELF, Revi A, Preston BL et al (2022) Climate change 2022: impacts, adaptation and vulnerability. In: Pörtner H-O, Roberts DC, Tignor M et al (eds) Contribution of working group II to the sixth assessment report of the intergovernmental panel on climate change. Cambridge University Press, Cambridge, UK and New York, NY, USA, pp 2655–2807
56. European State of the Climate (2020) Copernicus climate change service, Full report. In: European state of the climate 2020 summary, Copernicus climate change service
57. Teutschbein C, Quesada Montano B, Todorović A, Grabs T (2022) Streamflow droughts in Sweden: spatiotemporal patterns emerging from six decades of observations. J Hydrol Reg Studies 42:101171. https://doi.org/10.1016/j.ejrh.2022.101171
58. Beguería S, Vicente-Serrano SM, Reig F, Latorre B (2014) Standardized precipitation evapotranspiration index (SPEI) revisited: parameter fitting, evapotranspiration models, tools, datasets and drought monitoring. Int J Climatol 34:3001–3023. https://doi.org/10.1002/joc.3887
59. Seneviratne SI, Zhang X, Adnan M et al (2021) Weather and climate extreme events in a changing climate supplementary material. In: Masson-Delmotte V, Zhai P, Pirani A et al (eds) Climate change 2021: the physical science basis. Contribution of working group I to the sixth assessment report of the intergovernmental panel on climate change
60. Europen Environment Agency E (2022) Global and European temperatures. In: EEA
61. Outtena S, Stefan S (2021) Extreme wind projections over Europe from the Euro-CORDEX regional climate models. Weather Climate Extremes 33:100363. https://doi.org/10.1016/j.wace.2021.100363
62. Owen LE, Catto JL, Stephenson DB, Dunstone NJ (2021) Compound precipitation and wind extremes over Europe and their relationship to extratropical cyclones. Weather Climate Extremes 33:100342. https://doi.org/10.1016/j.wace.2021.100342
63. Javed H, Brambilla A, Strang M (2022) An Australian climate-based characterization of hygrothermal risks for buildings. Energy Build 265:112086. https://doi.org/10.1016/j.enbuild.2022.112086
64. Aggarwal C, Ge H, Defo M (2023) Assessing mould growth risk of wood-frame walls using partial least squares (PLS) regression considering climate model uncertainties. Build Environ 238:110374. https://doi.org/10.1016/j.buildenv.2023.110374
65. Shirmohammadi M, Leggate W, Redman A (2021) Effects of moisture ingress and egress on the performance and service life of mass timber products in buildings: a review. Constr Build Mater 290:123176. https://doi.org/10.1016/j.conbuildmat.2021.123176
66. Neta A, Levi Y, Morin E, Morin S (2023) Seasonal forecasting of pest population dynamics based on downscaled SEAS5 forecasts. Ecol Model 480:110326. https://doi.org/10.1016/j.ecolmodel.2023.110326
67. Jahangir E, Deck O, Masrouri F (2013) An analytical model of soil–structure interaction with swelling soils during droughts. Comput Geotech 54:16–32. https://doi.org/10.1016/j.compgeo.2013.05.009
68. Ebi KL, Capon A, Berry P et al (2021) Hot weather and heat extremes: health risks. Lancet 398:698–708. https://doi.org/10.1016/S0140-6736(21)01208-3
69. Jelle BP, Nilsen TN (2011) Comparison of accelerated climate ageing methods of polymer building materials by attenuated total reflectance Fourier transform infrared radiation spectroscopy. Constr Build Mater 25:2122–2132. https://doi.org/10.1016/j.conbuildmat.2010.11.020
70. Mehrabi Z, Donner S, Rios P et al (2019) Can we sustain success in reducing deaths to extreme weather in a hotter world? World Dev Perspect 14:100107. https://doi.org/10.1016/j.wdp.2019.02.018
71. Salamanca F, Georgescu M, Mahalov A et al (2014) Anthropogenic heating of the urban environment due to air conditioning. J Geophys Res: Atmos 119:5949–5965. https://doi.org/10.1002/2013JD021225

72. Taylor J, McLeod R, Petrou G et al (2023) Ten questions concerning residential overheating in Central and Northern Europe. Build Environ 234:110154. https://doi.org/10.1016/j.buildenv.2023.110154
73. Liyanage DR, Hewage K, Hussain SA et al (2024) Climate adaptation of existing buildings: a critical review on planning energy retrofit strategies for future climate. Renew Sustain Energy Rev 199:114476. https://doi.org/10.1016/j.rser.2024.114476
74. Steenbergen RDJM, Koster T, Geurts CPW (2012) The effect of climate change and natural variability on wind loading values for buildings. Build Environ 55:178–186. https://doi.org/10.1016/j.buildenv.2012.03.010
75. Pérez-Bella JM, Domínguez-Hernández J, López-Julián PL et al (2024) Closing the gap between traditional wind-driven rain studies and the performance-based design of building façades: case study of the Netherlands. Build Environ 248:111083. https://doi.org/10.1016/j.buildenv.2023.111083
76. Pérez-Bella JM, Domínguez-Hernández J, Ibarz-Montaner E et al (2024) Incorporating façade-specific climatic factors to improve the ISO 15927–3 characterisation of wind-driven rain spells: dutch and Spanish case studies. Dev Built Environ 17:100326. https://doi.org/10.1016/j.dibe.2024.100326
77. D'Ayala D, Aktas YD (2016) Moisture dynamics in the masonry fabric of historic buildings subjected to wind-driven rain and flooding. Build Environ 104:208–220. https://doi.org/10.1016/j.buildenv.2016.05.015
78. Ranta V, Aarikka-Stenroos L, Ritala P, Mäkinen SJ (2018) Exploring institutional drivers and barriers of the circular economy: a cross-regional comparison of China, the US, and Europe. Resour Conserv Recycl 135:70–82. https://doi.org/10.1016/j.resconrec.2017.08.017
79. Ghisellini P, Cialani C, Ulgiati S (2016) A review on circular economy: the expected transition to a balanced interplay of environmental and economic systems. J Cleaner Prod 114:11–32. https://doi.org/10.1016/j.jclepro.2015.09.007
80. EEA (2022) Building renovation: where circular economy and climate meet—European environment agency. Briefing no 12(2022):1–13
81. ISO 20887:2020 (2020) Sustainability in buildings and civil engineering works—Design for disassembly and adaptability—Principles, requirements and guidance published
82. European Parliament and the Council of the European Union (2018) DIRECTIVE (EU) 2018/844 of the European parliament and of the council of 30 May 2018 amending directive 2010/31/EU on the energy performance of buildings and directive 2012/27/EU on energy efficiency. Off J Eur Union 156:75–91
83. European Parliament and the Council of the European Union (2012) Directive 2012/27/EU of the European parliament and of the council of 25 October 2012 on energy efficiency, amending directives 2009/125/EC and 2010/30/EU and repealing directives 2004/8/EC and 2006/32/EC. Official J European Union pp 1–56
84. Agency EE, Castellari S, Zandersen M et al (2021) Nature-based solutions in Europe: policy, knowledge and practice for climate change adaptation and disaster risk reduction. Publications Office of the European Union, Luxembourg
85. Madureira S, Flores-Colen I, de Brito J, Pereira C (2017) Maintenance planning of facades in current buildings. Constr Build Mater 147:790–802. https://doi.org/10.1016/j.conbuildmat.2017.04.195
86. Silvestre JD, Castelo AMP, Silva JJBC et al (2020) Energy retrofitting of a buildings' envelope: assessment of the environmental, economic and energy (3E) performance of a cork-based thermal insulating rendering mortar. Energies. https://doi.org/10.3390/en13010143
87. Corrêa D, Flores-Colen I, Dinis Silvestre J et al (2020) Old buildings' façades: fieldwork and discussion of thermal retrofitting strategies in a Mediterranean climate. Designs. https://doi.org/10.3390/designs4040045
88. Silvestre JD, Castelo AMP, Silva JJBC et al (2019) Retrofitting a building's envelope: sustainability performance of ETICS with ICB or EPS. Appl Sci. https://doi.org/10.3390/app9071285

89. ESICC (2023) ESICC 2023-Abstracts Book-Energy efficiency, Structural integrity in historical and modern buildings face climate change and circularity
90. Andrade Santos R, Flores-Colen I, Simões N, Dinis Silvestre J (2023) Auto-responsive technologies on opaque facades: Worldwide climatic suitability under current and future weather conditions. J Build Eng 63:105498. https://doi.org/10.1016/j.jobe.2022.105498
91. Parracha JL, Borsoi G, Flores-Colen I et al (2021) Performance parameters of ETICS: correlating water resistance, bio-susceptibility and surface properties. Constr Build Mater 272:121956. https://doi.org/10.1016/j.conbuildmat.2020.121956
92. Parracha JL, Veiga R, Gomes MG et al (2024) Hygrothermal behaviour of external thermal insulation composite systems (ETICS) to withstand biological colonisation. J Build Eng 86:108932. https://doi.org/10.1016/j.jobe.2024.108932
93. BEWARE (2021–1-PT01-KA220-VET-000027997) (2024) Intellectual output: beware trends practicum with case studies
94. Ogut O, Bartolucci B, Parracha JL et al (2023) Energy poverty in Portugal, Italy, and Norway: awareness, short-term driving forces, and barriers in the built environment. IOP Conf Ser Earth Environ Sci 1176:12023. https://doi.org/10.1088/1755-1315/1176/1/012023

A Passivhaus Case Study. Design, Execution, Performance, and Economic Considerations

Belén Zurro García, Sara González Moreno, Jaime Moyano Villameriel, and José Manuel González Martín

Abstract The lurking climate-related challenges have fostered the adoption of measures to cut off energy consumption across the building sector, a main contributor to the overall greenhouse gas emissions. The *Passivhaus* (PH) standard has become a canonical benchmark across the EU for achieving the nearly-zero-energy building milestone. The foundation of the PH standard resides in achieving a heating demand below 15 $kWh/m^2 \cdot yr$, by minimising heat losses, while preserving an elevated indoor air quality. In this research, the design and execution of a non-residential building, according to the PH standard, is presented. The site features a bioclimatic design, and the Passive House Planning Package tool is utilised during the design and execution phases to estimate the thermal transmittance of the building envelope, quantify thermal bridges, and calculate energy demands, all of which complied with the PH criteria. Blower door tests are performed post-execution, and the resulting air renovation rates are consistent with an airtight structure, according to the PH standard. Budgetary considerations are also presented. A comparison between the expected and final expenditure provides a useful guideline for attaining better cost-effectiveness in future PH projects.

1 Introduction

Climate change constitutes one of the greatest challenges that humanity must address during the twenty-first century, since it directly threatens social welfare, world health, and global economy [1, 2]. The steadily increasing planet temperature has prompted nations to undersign the Paris Agreement. This worldwide commitment seeks to avoid a global temperature increase above 1.5 °C, compared to pre-industrial times, by reducing global greenhouse gas (GHG) emissions [3]. At European level, the EU Green Deal has been issued by the European Commission, an extensive and long-term package of measures, policies, and action plans to confront the upcoming

B. Zurro García · S. González Moreno · J. Moyano Villameriel (✉) · J. M. González Martín
Grupo de Investigación en Tecnología, Edificación, Construcción y Arquitectura (GITECA), University of Burgos, Burgos, Spain
e-mail: jmoyano@ubu.es

M. Torres González and C. Rubio Bellido (eds.), *Towards Low and Positive Energy Buildings*, Green Energy and Technology, https://doi.org/10.1007/978-3-031-70851-0_2

climate challenges and ensure a climate-neutral economy by 2050 [2, 4]. Within the framework of the EU Green Deal, specific policies that intend to cut off GHG emissions, in alignment with the Paris Agreement, have been outlined. The Fit for 55 initiative is a key element that sets an ambitious, yet fundamental, goal: reduction in GHG emissions by, at least 55%, by 2030, in comparison with 1990, and zero net emissions by 2050 [5, 6]. Although deleterious gas emissions originate from several sectors, buildings represent a significant share. Up to 36% of GHG emissions stem from buildings, of which 72% are accountable to households [5–7]. These GHG emissions data are directly correlated to energy consumption. Buildings represent 40% of the energy usage across the EU [8]. A closer insight into these data reveals that space and water heating comprise major sources of energy expenditure—64.4% and 14.5% of the household average consumption [8, 9]. However, an estimated 60% of the thermal power is employed to counterbalance heat losses, due to poor insulation, inefficient appliances, and absence of integrated indoor climate control systems [10]. In fact, statistics show that the EU building stock is relatively obsolete, and there is a generalised absence of modernised energy efficacy measures. More specifically, 35% of the stock is above 50 years old and 90% of the current buildings have been constructed before 1990, according to, if any, already phased-out energy regulations [11].

The toll building energy efficiency has on GHG emissions has promoted the Renovation Wave, exemplified by the 2010/2031/EU recast issued by the Energy Performance of Buildings Directive (EPBD), which states that all buildings constructed after 31st December 2020 should be nearly-zero-energy [11, 12]. A new concept has arisen, as a result of the updated regulations: the nearly-zero-energy building (NZEB) [13]. The NZEB is described as "a building that has a very high energy performance [and] the nearly zero or very low amount of energy required should be covered to a very significant extent by energy from renewable sources" [12, 13]. The emergence of the NZEB has subsequently caused changes in the local regulations, in the form of tighter thermal performance benchmarks or a higher share of renewable sources in the primary energy consumption.

The EPBD recast has translated into more tangible outcomes. Within the period from 2012 to 2016, the share of NZEBs in the EU construction market (including new construction and renovations) has increased from 14 to 20%, accounting for a total of 1.25 million buildings, mainly in France and Germany. The rise of NZEB constitutes a motor for the construction sector that typically suffers from low innovation and adaptation difficulties to emerging trends. Renovation projects already involve 1% of the current EU buildings stock, and new construction of NZEBs boosts employment in the sector. In the longer term, these investments will reduce energy poverty and enhance public health, hence easing potential burdens to the social security systems [7, 14]. Neglecting the energy efficiency criteria constitutes an inadequate inversion strategy, since the building in question will be deemed obsolete from the very start of the life cycle. The growing pressure to achieve compliance with the NZEB standards has fostered the emergence of several energy efficiency strategies, which are conventionally classified into passive and active measures [15].

Active measures. This concept refers to in situ energy production from renewable sources (e.g. photovoltaic, geothermal, or wind power) and efficient cooling and/or heating devices [15, 16].

Passive measures. Advanced architecture design techniques featuring optimised spatial distribution or enhanced thermal insulation integrate the repertoire of passive strategies [15, 16]. Several innovative thermal insulation solutions have been implemented and accomplished energy efficiency goals, such as novel building envelope technologies, mechanical ventilation equipped with heat recovery systems or highly air-tight structures [17, 18]. Despite the wide variety of methodologies to achieve energy efficiency, the current EU regulations fail to provide a specific quantitative framework that enables a systematic and reliable categorisation of NZEBs.

Although the EPBD defines generic guidelines to achieve the climate objectives, each State Member (SM) is responsible for deploying a specific roadmap [19]. The heterogeneity of climates, different financial situations, and disparity of local energy-related regulations and sources across the SMs contribute to a lack of consensus about the NZEB definition. For example, most local NZEB descriptions include a maximum primary energy demand, expressed in $kWh/m^2{\cdot}yr$, and a wide spectrum of values can be seen. As an example, Belgium stablishes that, both, newly constructed residential buildings should not exceed 20 $kW/m^2{\cdot}yr$, yet the analogous standard in Estonia is 132 $kW/m^2{\cdot}yr$ [11].

Given the widespread lack of consensus, the *Passivhaus* (PH) standard has been adopted as a canonical and internationally accepted benchmark for the NZEB [20, 21]. The PH is a performance-based standard that intends to achieve energy efficiency by reducing thermal losses, the main culprit of domestic energy consumption. The PH principles do not solely focus on thermal insulation but aim to maximise occupants' comfort. Excellent air quality is maintained, and inner temperature is modulated upon heating or cooling the incoming fresh outdoor air [20, 22, 23]. Key elements in the PH are the building orientation, air tightness, and high-performance envelope, characterised by an insulated frame and negligible thermal bridge coefficients [22, 23]. The structural features are to be accompanied by a mechanical ventilation system equipped with a heat recovery system, an essential element to ensure indoor air quality and climate comfort [24].

The first PH-based building was constructed in 1991 in Darmstadt (Germany) and this technology has become widespread ever since [25]. Whilst most passive houses are located in Europe, several examples can be found in America, Asia, and the South Pacific [22, 23, 25]. By July 2023, more than 39,600 buildings worldwide have been certified to adhere to the PH [23, 26].

The requirements to achieve a PH certification relate to the spectrum of inner temperatures, air tightness, and energy consumption. The latest version of the PH criteria, issued in January 2023, establishes the following key quantitative descriptors [23, 27, 28]:

- Annual heating or cooling demand below 15 $kWh/m^2{\cdot}yr$

- Primary annual energy demand (combined heating, cooling, appliances, and lightning consumption) below 120 kWh/m^2·yr, of which, at least, 60 kWh/m^2·yr, must be produced by renewable sources.
- Air tightness. Indoor air exchange rate below 0.6 h^{-1}, at a 50 Pa indoor–outdoor differential pressure.
- Indoor temperature must not exceed 25 °C for more than 10% of the year.
- Humidity. Excess humidity (12 g/kg air) must not occur for more than 20% of the year, should no active cooling be installed—10% of the year if an active cooling system is available. Relative humidity must not decrease below 30%, in any case.
- Interactive and user-friendly indoor climate monitoring and regulation platforms must be installed.

This chapter focuses on a comprehensive analysis of a building designed and constructed in accordance with PH standards. The examination encompasses a comparative assessment of anticipated and actual performance, in terms of energy consumption, thermal insulation, air tightness, and budgetary considerations. This comparative study enables to determine whether the building aligns with the established PH criteria in real conditions.

1.1 Building Description

A projected single-storey annex of an adult daycare centre has been chosen for the current study. The venue is located in Villadiego municipality, in the Northern Province of Burgos (Spain). The area is governed by an oceanic climate or *Cfb* (temperate), according to the Köppen classification system—relatively mild summers and cold and humid winter periods [29]. The yearly average temperature oscillates around 10 °C, reaching an average maximum of 27 °C during the warmest period of the year (July and August), whereas the average minimum decreases down to –1 °C in the coldest month (January). The total rainfall is 550 mm/yr on average, autumn being the rainiest period [30, 31].

Emmepolis Novecento architecture study has designed and constructed a building according to the PH standards conceived to be fully accessible for mobility-impaired persons, since it will serve as annex of the pre-existing adult daycare facility (Fig. 1). The trapezoid-shaped floor plan comprises a surface of 343.63 m^2 and shares one side with the adjacent building.

1.2 Assessment of Passive Measures

The *Passivhaus* Inistitute (PHI) provides a design and planning tool, the *Passivhaus* Planning Package (PHPP), typically in the form of a spreadsheet, to calculate heating and cooling demands and total energy consumption, according to the building design;

Fig. 1 Exterior view of the South (right) and Western (left) façades

envelope; heating, ventilation, and air conditioning specifications; and climate conditions [20, 22, 23, 32]. Based on the input data, the PHPP determines whether the building will adjust to the PH standards. The PHPP is applied prior the execution phase to verify the suitability of the proposed design [32].

To ensure compliance with the PH standards, actions were undertaken according to a variety of criteria, listed below.

1.3 Thermal Insulation

As illustrated by the profile building plan, the thermal envelope (represented by the coloured segments in Fig. 2) engulfs the entire inner space in a continuous fashion, including the foundation. The divisor wall between the new facility and the pre-existing adult daycare centre is also insulated, to achieve an indoor climate and energy independence from the adjacent building. The thermal insulation layer is composed of either expandable polystyrene with graphitic carbon (*Neopor*®) or extruded polystyrene (*Topox*) [33–35].

1.4 Thermal Bridge Reduction

According to the Spanish Technical Building Code (CTE), thermal bridges are defined as sections in the building's thermal envelope, wherein there is a disruption, caused by a transition in the structure enclosure or a sudden differences in thermal transmittance [36]. Junctions between building elements, such as roofing-beam,

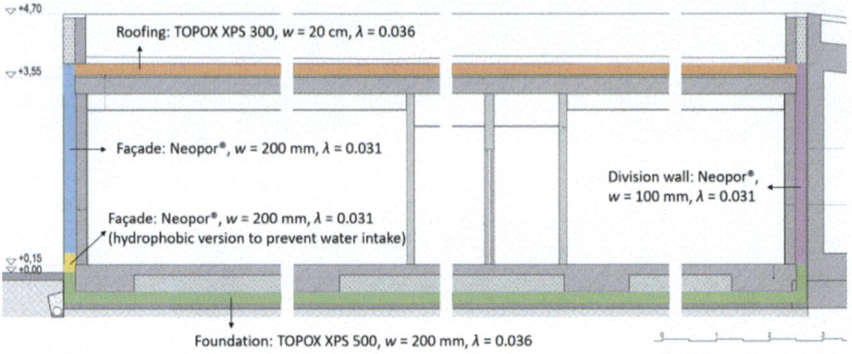

Fig. 2 Profile building plan. The thermal envelope characteristics, for each segment, are presented in the form of material, width (w) and thermal conductivity (λ)

pillars or window frames, constitute typical thermal discontinuities and are detrimental towards energy efficiency. To minimise thermal bridges, it is crucial to ensure a non-interrupted thermal envelope. The so-called pencil rule is typically employed during the design phase to prevent thermal bridges (*i.e.*, should the roof insulation be properly connected to that of the walls, and ultimately to the floor, it ought to be possible to draw a continuous line on the section plan, following the insulation layer, without lifting the pencil form the paper) [37].

A quantitative evaluation of the thermal bridges has been performed via the PHPP, based on UNE-EN ISO 10077–1:2020 standard, and the values are presented in Table 1 [38].

Most thermal bridges display minimum linear thermal transmittance, thanks to the continuous thermal insulation. The junction between the façade and pillars deviates, though, from the trend, as the transmittance is exceptionally high. By visualising the pillar-façade junction cross-section, it is appreciated that the thermoclay layer is interrupted by the pillar, thus inducing a heterogenicity in the thermal envelope (Fig. 3a). For future reference, the floor plan could be configured so the pillars are adjacent to—not embedded—the thermoclay layer (Fig. 3b).

The thermal insulation layers are incorporated to the remaining elements that integrate the building enclosure. According to standard UNE-EN ISO 6946:2021 and

Table 1 Thermal bridge assessment

Thermal bridge	Length (m)	Linear thermal transmittance (W/m·K)
Foundation-terrain	0.00	− 0.053
Foundation-division wall	7.15	0.093
Roofing-division wall	75.57	0.124
Access door-terrain	7.13	0.106
Pillars-façade	22.00	0.500

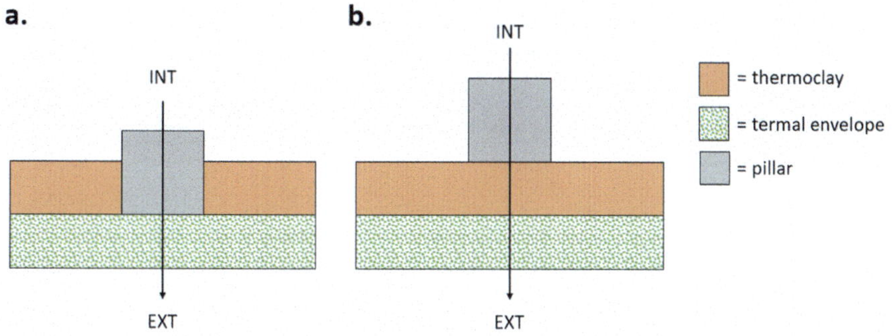

Fig. 3 Pillar-façade junction cross-section. **a** Actual cross-section. **b** Suggested cross-section layout to reduce the thermal bridge

Table 2 Thermal transmittances for the four envelope segments and comparison with the CTE standards

Location	Thermal transmittances (W/m^2·K)		
	Calculated	CTE maximum	CTE recommendation
Façade	0.136	0.37	0.23
Roofing	0.167	0.33	0.19
Floor	0.164	0.59	0.48
Division wall	0.134	0.59	0.48

the enclosure components specification data sheets, the theoretical thermal transmittances for each envelope segment can be calculated [36]. Table 2 indicates that the thermal transmittances along the four segments are, at least, 50% below the maximum value, permitted by the CTE, given Villadiego's climate (E-class, in the code nomenclature). Additionally, the CTE sets recommended values, and calculations that are likewise below the latter benchmark, which demonstrates the thermal envelope features largely compensate for thermal bridges.

1.5 High-Performance Carpentry and Window Glazing

Carpentry and window glass features largely influence thermal performance. Window glazing and frame elements of similar thermal transmittance than the overall envelope will be selected to minimise thermal bridges. Parameters such as thermal transmittance (U) and solar energy transmittance coefficient (G) constitute the principal selection criteria (Table 3).

The larger solar energy inputs in the southern and eastern façades justify the choice of higher solar energy transmittance and lower thermal transmittance glazing,

Table 3 Glazing and carpentry features based on orientation

Elements	Orientation	Specifications	Area (m^2)	G (kWh/yr)	U ($W/m^2{\cdot}K$)
Carpentry	North	Polyvinylchloride	25.65	1,240	1.035
	East		45.76	10,563	0.749
	South		6.89	441	1.048
	West		22.00	2,387	0.742
Glazing	South and West	PlaniClear-PlanitTherm XN F3 F5 3 + 3 / 18 argon/ 4 / 18 argon / 3 + 3	71.70	0.53	0.54
	North	Lamiglass 4 + 4 GuardianSun / 16 argon/ 4 / 16 argon/ 3 + 3	25.69	0.38	0.84

compared to the North-facing elements. The building design seeks to prevent overheating during the summer months, while maximising natural lightning and thermal gains during the winter.

1.6 Air Tightness

The overall structure should prevent air infiltrations that constitute heat losses and detrimentally affect indoor comfort. In addition, the airtight frame also provides an excellent noise insulation. Critical elements that may lead to air leakage are the junctions between walls and window frames. A precise and detail-oriented installation of an air-tight membrane is necessary to reinforce these weak points. The PHI considers a building to be airtight, should the number of hourly air renovations not exceed 0.6 h^{-1}. Air tightness will be assessed by the blower door test. The airtightness of renovated facilities according to the PH standard has been measured by the blower door test on several occasions [37, 38]. The test commences by the temporary sealing of all the structure's openings. A door, or a window, fitted with a blower fan is thereby installed and generates a pressure gradient, typically 50 Pa, between the inside and outside of the building. The amount of leaked air during the individual pressurisation and depressurisations tests is directly translated into the averaged renovation rate [39].

The sole access door potentially constitutes the primary source of air leakage—a structural weak point—since it is levelled to the interior and exterior grounds for enhanced accessibility reasons. Hence, the blower fan device is installed in a window to acquire a more representative insight on the structure's air hermeticity (Fig. 4a, b). The test performed at 50% execution progress afforded an averaged air renovation rate of 0.27 h^{-1} (Table 4). Despite some window sealing being unproperly installed, as discovered by a subsequent inspection (Fig. 4c), the air tightness is less than 50% of the maximum. The sealing was properly placed afterwards (Fig. 4d).

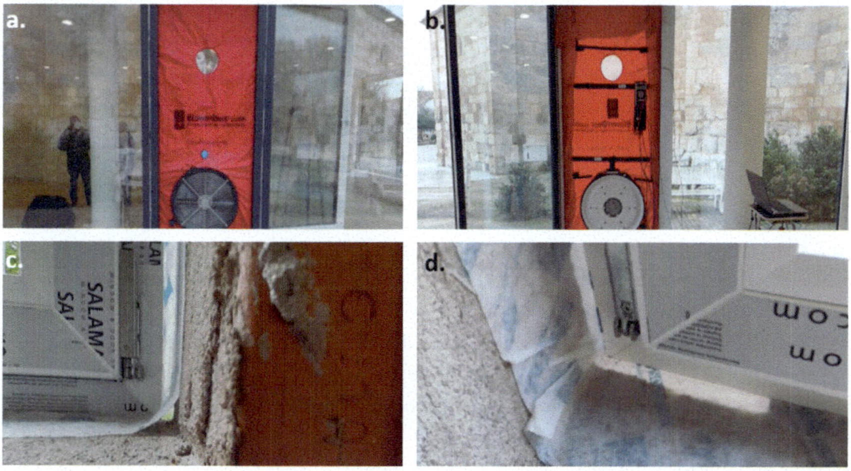

Fig. 4 **a** Outside view of window-mounted blower door device. **b** Inside view. **c** The misplacement of the sealing tape creates a void at the window frame corners. **d** Correct placement of the tape

Table 4 Blower door test results. All datapoints correspond to a 50 Pa differential pressure

Execution progress	Date	Air changes (h^{-1})		
		Depressurisation	Pressurisation	Average
50%	28/09/2020	0.27	0.27	0.27
100%	14/01/2021	0.50	0.57	0.54

Upon completion, the air tightness has decreased—0.54 h^{-1}—yet the facility still complies with the PH standard. The boring associated to the installation of water and electricity supplies is responsible for the loss of air hermeticity.

1.7 MVHR (Mechanical Ventilation with Heat Recovery) System

A healthy indoor atmosphere—low concentration of CO_2 (below 1200 ppm) [24], absence of condensation—requires from the steady action of a ventilation device. A MVHR system creates a double airflow that enables a nearly quantitative heat exchange between the incoming fresh outdoor air and the exhaust, keeping the indoor temperature constant and minimising heat losses. [40] The MVHR will be additionally equipped with air filters, to retain potential outdoor pollutants (*e.g.* dust or pollen) and indoor humidity, and CO_2 sensors will be installed to monitor air quality.

The ventilation system is a Zehnder, ComfoAir XL 200 apparatus, capable of providing a ventilation rate from 750 to 1500 m^3/h, adequate for the volume of the facility, 772 m^3. The heat recovery system comes up to a 75% performance threshold, in alignment with the European regulations [41].

1.8 Bioclimatic Architecture

The PH design has, a priori, been conceived to be virtually applicable to any climate and geographical location [25]. Evaluation of PH performance across the globe demonstrates that thermal comfort is indeed easily achieved and maintained during winter periods; however, challenges are encountered throughout the warmest months [42–44]. Empirical evidence showcases that PH-designed buildings, and other NZEBs in general, are prone to overheating (conventionally accepted to occur when the indoor temperature surpasses 25 °C, and ideally, such a situation should not account for longer than 10% of the year) during summer periods, causing discomfort to the occupants [23, 44, 45]. Such an issue is indeed reflected in the biased geographical distribution of PH buildings across the EU. A majority is located in Northern and Central European countries, whereas warmer Mediterranean countries have not experienced a comparable success of this technology [26, 46]. During the warmest months, the thermal insulation, window glazing, and air tightness become enemies of thermal comfort [44]. Additional detrimental factors are non-optimal building orientation, lack of shading, and insufficient ventilation [43]. Recurring overheating could potentially encourage the installation of active cooling devices, of which consumption may override the energy saving afforded by the passive elements [42, 44, 45].

Overcoming this challenge requires from a holistic approach that includes the simultaneous implementation of passive measures tailored to the climate conditions of the building location: a bioclimatic design [23, 25, 47]. In the specific case of the PH building in Villadiego, the annual average sunshine duration at the venue exceeds 2,000 h—the impact of solar energy on the building's performance should thus not be underestimated. The following interventions have been undertaken accordingly:

- Compact building shape. Building compactness is defined as the ratio between the thermal envelope surface area and the enclosed volume [48]. Building compactness has traditionally been a recurring architectural solution in Mediterranean countries to avoid overheating [46].
- The external surface is 349.63 m^2 and the enclosed volume 772 m^3, thereby resulting in a compactness ratio of 0.45 m^{-1}. The Passivhaus Primer—Designer's guide recommends a value below 0.7 m^{-1}, and thus, the facility's design is optimal according to this criterion [48].
- North–South orientation. The amplest façade is South-oriented to maximise solar heat gains during the winter, whilst the Northern façade will feature less surface to minimise heat losses. The dual North–South orientation creates a zero-energy consuming natural air drift that provides cooling during the warmest months [49].

- Window sizes. Windows oriented to the South are conceived to maximise solar radiation gains—the glazing spans from 0.15 m above the floor up to the ceiling level (2.20 m total height). The Northern-facing elements are 1.50 m high, to reduce heat losses during the winter.

Despite these measures, the PHPP-based calculations estimate that overheating shall occur for 23% of the year, well above the recommendations [23]. An additional level of protection towards overheating, not considered by the PHPP at this stage, will be provided by solar admission controls. Windows will be equipped with motorised roller blinds to on-demand block excess solar radiation inputs.

As future research, an evaluation of the actual overheating frequency should be conducted to corroborate the accuracy of the PHPP estimates and to determine the influence of the roller blinds and the occupants' habits.

1.9 Energy Demand

The combined building features and passive elements (thermal bridge minimisation, thermal envelope, bioclimatic design, and air tightness) are expected to meet the heating, cooling, and primary energy demands determined by the PHI. The PHPP enables to calculate the energy demands, based various input data, such as the window specifications, blower door test results, estimated occupancy level and façade, roofing, and floor plan surface areas. The calculations depicted in Table 5 show that compliance with the PHI standards is indeed achieved for these three categories.

The heating demand stems from the difference between the solar energy inputs and the thermal losses. The latter are calculated based on the expected capacity (40 people), air exchange rate, thermal transmittances of the construction elements, and the indoor–outdoor temperature gradient, whereas the former are derived from climate statistics. To meet the heating demands, the building will be connected to the districting heating network of Villadiego and equipped with the subsequent substation (54 kW).

After one year in service as adult daycare centre, monthly heating consumption data were compiled, and the actual demand per surface unit is calculated and

Table 5 Estimated energy demands

Energy demand category	Losses by ventilation & transmittance (kWh/m^2·yr)	Solar radiation inputs (kWh/m^2·yr)	Annual demand (kWh/m^2·yr)	PHI maximum (kWh/m^2·yr)	Compliant?
Primary	–	–	74.04	120	Yes
Heating	68.4	54.7	13.64	15	Yes
Cooling	–	–	9.90	15	Yes

compared to the foreseen value (Table 6). The accumulated yearly heating consumption has been 6,523 kWh/yr, being the winter months accountable for a significant share. Considering the facility usable surface —343.63 m^2—the surface specific yearly heating demand exceeds the PHI standard by 27%.

Although the energy consumption falls out from the PH standard, in the context of average energy demands of non-residential buildings, the described facility comes up to a rather ambitious benchmark. Data from various studies concerning office and educational buildings indicate the normal consumption intervals range between 80 and 140 kWh/m^2·yr, whereas for daycare centres in humid continental climates, of which energy needs are comparable to the present case, the average is 150 kWh/m^2·yr [50–56].

The unexpected deviation from the optimal heating consumption opens the door for future investigations about the facility's energy performance. Preliminary speculations pinpoint the daycare centre staff, and users may not be fully aware of the installed energy-saving measures and may not take the necessary actions to achieve peak performance. Unnecessary window opening or tilting, due to unawareness that the MVHR system suffices for air renovation, or setting and excessively high indoor temperature, given the users tendency to feel thermal discomfort are plausible sub-optimal practices.

In fact, literature evidence highlights that performance of energy measures is largely conditioned by the user's indoor habits, good practices, and notions on energy efficiency [44]. A starting point for future research may concern studying how the building staff operates the user-accessible energy-related measures (*e.g.*, window opening frequency, indoor temperature, and MHVR settings). Should the operation of the building deviate from the guidelines, the staff could be instructed to follow

Table 6 Actual heating consumption

Month/ year	Heating consumption (kWh)	Yearly consumption (kWh)	Surface specific yearly consumption (kWh/m^2·yr)	Compliant?	Deviation
12/22	735	6523	18.98	No	+ 26.6%
01/23	939				
02/23	725				
03/23	597				
04/23	500				
05/23	492				
06/23	484				
07/23	200				
08/23	239				
09/23	240				
10/23	580				
11/23	792				

Table 7 Budgeted and actual execution costs

Item	Planned cost (€)[a, b]	Actual cost (€)[b]	Over cost (%)
Envelope	12,030.98	21,622.24	79.7
Flooring	8,329.38	25,440.04	205.4
Carpentry and window glazing	27,077.49	29,722.45	9.8
Electrical system	13,971.03	20,236.11	44.8

[a] According to PHPP. [b] Includes 21% VAT

the recommendations for a prolonged period and monitor the subsequent heating consumption.

1.10 Budget

The execution of the facility has encountered budget overruns, primarily associated to various items, mentioned in Table 7.

Installation of the envelope components required from more extensive workforce than initially expected. The precise and detail-oriented application of plaster and hermetic paint layers is, generally, laborious and time-consuming. Similarly, the meticulous work demanded for the installation of the air-tight sealing tape around the window carpentries caused overruns, yet to a lesser extent. To ensure a more efficient execution in the future, PH-specialised construction staff may be hired.

The installation of a costlier interruption-free linoleum flooring and non-projected outdoor sidewalk refurbishment, to enhance accessibility, accounts for the excess expenditure in this item. The electrical system needed adjustments during execution to ensure compatibility with the district heating substation, plus a distribution board, of which cost was not initially considered.

2 Conclusions

The upcoming climate—and environment-related changes have impelled public authorities to deploy energy and decarbonisation policies targeted at the construction and building sector, exemplified by the widespread implementation of the NZEB concept across the EU.

Several technologies have thus emerged to ensure alignment with the updated environmental framework. The PH standard stands out as an ambitious, yet realistic, benchmark for NZEBs. Introduced and regulated by the PHI, the PH certification comprises a series of interventions with the sole objective of reducing primary energy consumption in dwellings and non-residential buildings. A PH building features a highly efficient envelope, characterised by low thermal transmittance materials (including window glazing and carpentry), suppression of thermal bridges and high air tightness. Despite the apparent structural hermeticity, the PH guarantees high indoor air quality via the constant air renewal without heat losses, promoted by a MVHR device.

In this research, the features of a new-construction PH building, intended to serve as an adult daycare centre, have been described. The facility has been crafted according to the bioclimatic design principles: ample exposure to the South enables substantial solar energy inputs, while a natural air drift between the Northern and Southern façades provides a natural source of cooling.

The building performance was examined via the PHPP tool and post-execution measurements.

- Air tightness. The state-of-the-art blower door test was employed to assess the air hermeticity of the envelope. The air tightness measurements conducted during and post-execution were satisfactory, since air renovation rates of 0.27 and 0.54 h^{-1} were afforded, respectively, which comply with the PH guidelines.
- Thermal envelope performance. PHPP-based calculations show that the envelope's thermal transmittances are well below the CTE-dictated maxima and even the recommended values. Thermal bridges were also quantified and do not generally exceed 0.125 $W/m^2{\cdot}K$, except for the pillar-façade junction (0.500 $W/m^2{\cdot}K$). The latter constitutes a potential element where to direct future improvements.
- Energy demands. Based on the envelope's thermal transmittances, climate statistics, and building capacity, the annual heating demand is estimated to be below the PHI maximum. However, after one year in service, the heating consumption exceeds the standard by nearly 27%. It is speculated that the occupants' habits may be negatively affecting energy performance, yet a posterior study should be conducted to further delve into the origin of this deviation.
- Budget. The actual execution cost has overrun the PHPP-estimated budget. The over costs mainly account for the extra workforce needed to correctly install the air hermeticity elements. To minimise budget overruns in future PH projects, it is suggested that construction staff experienced in air-tight facilities should be incorporated during the execution phase.

Ultimately, the presented research will serve as a reference point for future NZEB-related actions, specially within the context of PH projects, by providing guidelines to further enhance building energy efficiency and guarantee a more cost-effective passive measure implementation.

References

1. Hellin J, Fisher E (2019) The achilles heel of climate-smart agriculture. Nat Clim Change 9(7):493–494. https://doi.org/10.1038/s41558-019-0515-8

2. Tsemekidi Tzeiranaki S, Bertoldi P, Economidou M, Clementi EL, Gonzalez-Torres M (2023) Determinants of energy consumption in the tertiary sector: Evidence at European level. Energy Rep 9:5125–5143. https://doi.org/10.1016/j.egyr.2023.03.122
3. Garaffa R et al (2023) Stocktake of G20 countries' climate pledges reveals limited macroeconomic costs and employment shifts. One Earth 6(11):1591–1604. https://doi.org/10.1016/j.oneear.2023.10.012
4. European Commission (2019) The European green deal, Brussels
5. European Commission and Ipsos Belgium (2019) Comprehensive study of building energy renovation activities and the uptake of nearly zero-energy buildings in the EU
6. Buckley N, Mills G, Reinhart C, Berzolla ZM (2021) Using urban building energy modelling (UBEM) to support the new European Union's Green Deal: Case study of Dublin Ireland. Energy Build 247(111115):1–11. https://doi.org/10.1016/j.enbuild.2021.111115
7. European Commission (2021) Making our homes and buildings fit for a greener future, https://doi.org/10.2775/081949
8. Eurostat (2023) Energy statistics—an overview. Accessed: Nov. 27, 2023. [Online]. Available: https://ec.europa.eu/eurostat/statistics-explained/index.php?title=Energy_statistics_-_an_overview#Final_energy_consumption
9. Aneli S, Arena R, Marco Tina G, Gagliano A (2023) Improvement of energy self-sufficiency in residential buildings by using solar-assisted heat pumps and thermal and electrical storage. Sustain Energy Technol Assess 60(103446):1–9. https://doi.org/10.1016/j.seta.2023.103446
10. Arregi B, Garay-Martinez R, Astudillo J, García M, Ramos JC (2020) Experimental and numerical thermal performance assessment of a multi-layer building envelope component made of biocomposite materials. Energy Build 214(109846):1–10. https://doi.org/10.1016/j.enbuild.2020.109846
11. D'Agostino D, Tzeiranaki ST, Zangheri P, Bertoldi P (2021) Assessing nearly zero energy buildings (NZEBs) development in Europe. Energ Strat Rev 36(100680):1–17. https://doi.org/10.1016/j.esr.2021.100680
12. European Parliament (2010) Directive 2010/31/EU of the European Parliament and of the Council, Brussels
13. European Commission (2016) Commission Recommendation (EU) 2016/1318 on guidelines for the promotion of nearly zero-energy buildings and best practices to ensure that, by 2020, all new buildings are nearly zero-energy buildings, [Online]. Available: http://www.sciencedirect.com/science/article/pii/S0306261911007811/
14. BPIE and Renovate Europe (2020) Building Renovations: A kick-starter for the EU recovery
15. Oh J, Hong T, Kim H, An J, Jeong K, Koo C (2017) Advanced strategies for net-zero energy building: Focused on the early phase and usage phase of a building's life cycle. Sustain Switz 9(12):1–52. https://doi.org/10.3390/su9122272
16. Economidou M, Todeschi V, Bertoldi P, D'Agostino D, Zangheri P, Castellazzi L (2020) Review of 50 years of EU energy efficiency policies for buildings. Energy Build 225(110322):1–20. https://doi.org/10.1016/j.enbuild.2020.110322
17. Aslani A, Bakhtiar A, Akbarzadeh MH (2019) Energy-efficiency technologies in the building envelope: life cycle and adaptation assessment. J Build Eng 21:55–63. https://doi.org/10.1016/j.jobe.2018.09.014
18. Blázquez T, Ferrari S, Suárez R, Sendra JJ (2019) Adaptive approach-based assessment of a heritage residential complex in southern Spain for improving comfort and energy efficiency through passive strategies: a study based on a monitored flat. Energy 181:504–520. https://doi.org/10.1016/j.energy.2019.05.160
19. Zangheri P, D'Agostino D, Armani R, Bertoldi P (2022) Review of the cost-optimal methodology implementation in member states in compliance with the energy performance of buildings directive. Buildings 12(9):1–19. https://doi.org/10.3390/buildings12091482
20. Dequaire X (2012) Passivhaus as a low-energy building standard: contribution to a typology. Energy Effic 5(3):377–391. https://doi.org/10.1007/s12053-011-9140-8

21. Rodríguez-Soria B, Domínguez-Hernández J, Pérez-Bella JM, del Coz-Díaz JJ (2014) Review of international regulations governing the thermal insulation requirements of residential buildings and the harmonization of envelope energy loss. Renew Sustain Energy Rev 34:78–90. https://doi.org/10.1016/j.rser.2014.03.009
22. Johnston D, Siddall M, Ottinger O, Peper S, Feist W (2020) Are the energy savings of the passive house standard reliable? A review of the as-built thermal and space heating performance of passive house dwellings from 1990 to 2018. Energy Effic 13(8):1605–1631. https://doi.org/10.1007/s12053-020-09855-7
23. Schnieders J et al (2020) Design and realisation of the passive house concept in different climate zones. Energy Effic 13(8):1561–1604. https://doi.org/10.1007/s12053-019-09819-6
24. Guillén-Lambea S, Rodríguez-Soria B, Marín JM (2016) Review of European ventilation strategies to meet the cooling and heating demands of nearly zero energy buildings (nZEB)/ Passivhaus. Comparison with the USA. Renew Sustain Energy Rev 62:561–574. https://doi.org/10.1016/j.rser.2016.05.021
25. Schnieders J, Feist W, Rongen L (2015) Passive Houses for different climate zones. Energy Build 105:71–87. https://doi.org/10.1016/j.enbuild.2015.07.032
26. Passive House Institute (2023) Passive house certifications worldwide, Darmstadt
27. Passive House Institute (2023) Criteria for buildings, Darmstadt, [Online]. Available: www.passivehouse.com
28. Yakimchuk T, Linhares P, Hermo V (2023) Evaluation of a modular construction system in accordance with the Passivhaus standard for components. J Build Eng 76(107234):1–24. https://doi.org/10.1016/j.jobe.2023.107234
29. Peel MC, Finlayson BL, Mcmahon TA (2007) Hydrology and earth system sciences updated world map of the Köppen-Geiger climate classification. Hydrol Earth Syst Sci 11:1633–1644
30. Meteo Illes Balears (2023) Clasificación de Köppen en territorio español. Caracterización y cambios recientes. Accessed: Nov. 25, [Online]. Available: http://www.meteoillesbalears.com/?p=272
31. AEMET (2023) Guía Resumida del Clima en España (1981–2010), https://doi.org/10.31978/281-1mete2-011-4.
32. Marschall MG, Sepulveda Corradini JP (2023) An industry case study using parametric modelling to facilitate passive house design. Int J Architect Comput 21(3):421–444. https://doi.org/10.1177/14780771231176140
33. BASF (2023) Neopor. Accessed: Nov. 28, [Online]. Available: https://neopor.basf.us/
34. EFYOS (2016) Products catalogue, [Online]. Available: www.efyos.es
35. Abina A, Puc U, Jeglič A, Zidanšek A (2016) Structural characterization of thermal building insulation materials using terahertz spectroscopy and terahertz pulsed imaging. NDT E Int 77:11–18. https://doi.org/10.1016/j.ndteint.2015.09.004
36. Asociación Española de Normalización (2021) UNE-EN ISO 6946:2021—Componentes y elementos para la edificación. Resistencia térmica y transmitancia térmica. Método de cálculo
37. Zurro García B, González Moreno S, González Martín JM and Paredes Núñez AM (2023) The principles of the passivhaus standard applied to rehabilitation. Case study, In: Building engineering facing the challenges of the 21st Century, vol 345, D. Bienvenido-Huertas and Durán-Álvarez J, eds, in Lecture Notes in Civil Engineering, vol 345, Singapore: Springer Nature Singapore, pp 473–484. https://doi.org/10.1007/978-981-99-2714-2
38. Echarri-Iribarren V, Sotos-Solano C, Espinosa-Fernández A, Prado-Govea R (2019) The Passivhaus standard in the Spanish Mediterranean: evaluation of a house's thermal behaviour of enclosures and airtightness. Sustainability 11(13):1–25. https://doi.org/10.3390/su11133732
39. Hsu YS, Zheng X, Cooper E, Gillott M, Wood CJ (2021) Evaluation of the indoor pressure distribution during building airtightness tests using the pulse and blower door methods. Build Environ 195(107742):1–12. https://doi.org/10.1016/j.buildenv.2021.107742
40. Guerra Santin O, Grave A, Jiang S, Tweed C, Mohammadi M (2021) Monitoring the performance of a Passivhaus care home: lessons for user-centric design. J Build Eng 43(102565):1–14. https://doi.org/10.1016/j.jobe.2021.102565

41. European Commission (2014) Commission regulation (EU) No 1253/2014 of 7 July 2014 implementing Directive 2009/125/EC of the European Parliament and of the Council with regard to ecodesign requirements for ventilation units
42. Figueiredo A, Kämpf J, Vicente R (2016) Passive house optimization for Portugal: overheating evaluation and energy performance. Energy Build 118:181–196. https://doi.org/10.1016/j.enbuild.2016.02.034
43. Mlecnik E, Schütze T, Jansen SJT, De Vries G, Visscher HJ, Van Hal A (2012) End-user experiences in nearly zero-energy houses. Energy Build 49:471–478. https://doi.org/10.1016/j.enbuild.2012.02.045
44. McLeod RS, Hopfe CJ, Kwan A (2013) An investigation into future performance and overheating risks in Passivhaus dwellings. Build Environ 70:189–209. https://doi.org/10.1016/j.buildenv.2013.08.024
45. Colclough S, Kinnane O, Hewitt N, Griffiths P (2018) Investigation of nZEB social housing built to the Passive House standard. Energy Build 179:344–359. https://doi.org/10.1016/j.enbuild.2018.06.069
46. Borrallo-Jiménez M, LopezDeAsiain M, Esquivias PM, Delgado-Trujillo D (2022) Comparative study between the Passive House Standard in warm climates and Nearly zero energy buildings under Spanish technical building code in a dwelling design in Seville, Spain. Energy Build 254(111570):1–17. https://doi.org/10.1016/j.enbuild.2021.111570
47. Figueroa-Lopez A, Arias A, Oregi X, Rodríguez I (2021) Evaluation of passive strategies, natural ventilation and shading systems, to reduce overheating risk in a passive house tower in the north of Spain during the warm season. J Build Eng 43(102607):1–15. https://doi.org/10.1016/j.jobe.2021.102607
48. McLeod R, Mead K and Standsen M (2010) Passivhaus primer: Designer's guide
49. Manzano-Agugliaro F, Montoya FG, Sabio-Ortega A, García-Cruz A (2015) Review of bioclimatic architecture strategies for achieving thermal comfort. Renew Sustain Energy Rev 49:736–755. https://doi.org/10.1016/j.rser.2015.04.095
50. Ruusala A, Laukkarinen A, Vinja J (2009) Energy consumption of Finnish schools and daycare centers and the correlation to regulatory building permit values. Energy Policy 119:183–195
51. Pietrapertosa F et al (2021) An educational awareness program to reduce energy consumption in schools. J Clean Prod 278(123949):1–13. https://doi.org/10.1016/j.jclepro.2020.123949
52. Hee Chung M, Eun Kim S, Woo Song Y, Chul Park J (2023) Strategies for improving impact of energy renovation: a case study on Korean daycare centers. Energy Build 284(112844):1–12. https://doi.org/10.1016/j.enbuild.2023.112844
53. Xia J et al (2014) Comparison of building energy use data between the United States and China. Energy Build 78:165–175. https://doi.org/10.1016/j.enbuild.2014.04.031
54. Nunes P, Lerer MM, Carrilho Da Graça G (2013) Energy certification of existing office buildings: Analysis of two case studies and qualitative reflection. Sustain Cities Soc 9:81–95. https://doi.org/10.1016/j.scs.2013.03.003
55. Zhao J, Wu Y, Zhu N (2009) Implementing effect of energy efficiency supervision system for government office buildings and large-scale public buildings in China. Energy Policy 37:2019–2086
56. Boyano A, Hernandez P, Wolf O (2013) Energy demands and potential savings in European office buildings: case studies based on EnergyPlus simulations. Energy Build 65:19–28. https://doi.org/10.1016/j.enbuild.2013.05.039

Techno-Economic Assessment of Energy Efficiency in a Building at the Technical University of Manabí (Portoviejo, Ecuador)

Jorge Milton Velepucha Sánchez, Ramona Panchana, Ángel Rafael Arteaga Linzan, Jose Gregorio Zevallos Cobeña, and Daniel Delgado

Abstract The techno-economic assessment of energy efficiency in a university teaching building was conducted to analyse the feasibility of investments in energy optimization improvements within this educational environment. This evaluation focused not only on reducing energy consumption but also on optimizing available energy resources. The methodology involved an energy analysis, performance simulations, and a detailed economic analysis, considering initial, operational costs, and benefits throughout the building's lifecycle. The results aim to identify whether the implementation of efficient lighting systems, adequate insulation, and renewable energy sources can generate significant economic savings and a notable reduction in energy consumption in the teaching building. Additionally, the optimization of climate control systems was highlighted as a key opportunity, and the positive influence of incentive policies and financing in promoting energy efficiency in educational settings was acknowledged. It was concluded that investments in energy efficiency are beneficial both economically and environmentally for university buildings, contributing to long-term sustainability and the reduction of operational costs.

J. M. Velepucha Sánchez · Á. R. Arteaga Linzan
Departamento de Mecánica, Facultad de Ingeniería y Ciencias Aplicadas, Universidad Técnica de Manabí, Avenida José María Urbina, Portoviejo EC130105, Manabí, Ecuador

R. Panchana · D. Delgado (✉)
Departamento de Construcciones Civiles, Arquitectura y Geología, Facultad de Ingeniería y Ciencias Aplicadas, Universidad Técnica de Manabí, Avenida José María Urbina, Portoviejo EC130105, Manabí, Ecuador
e-mail: daniel.delgado@utm.edu.ec

J. G. Zevallos Cobeña
Red de Desarrollo Urbano Sostenible de Manabí, Portoviejo, Ecuador

M. Torres González and C. Rubio Bellido (eds.), *Towards Low and Positive Energy Buildings*, Green Energy and Technology, https://doi.org/10.1007/978-3-031-70851-0_3

1 Introduction

Energy is a multifaceted concept spanning various scientific disciplines and is essential for the development and functioning of society. The ability to perform work or induce changes in a system is governed by the principle of energy conservation, enabling its transformation into various forms such as thermal, luminous, kinetic, and chemical [1]. Humanity's progress has been closely linked to its capacity to manipulate and harness energy, from the use of fire to the advanced technologies of the modern era [2].

Efficient energy management is crucial on a global scale for a sustainable future. By optimizing consumption to achieve maximum performance with minimal resource use, energy efficiency plays a vital role in mitigating environmental impacts and combating climate change [3]. Sustainable construction and the optimization of building systems such as heating, ventilation, and lighting are fundamental to reduce operational costs and promote healthy environments [4].

Over 100 countries have set net-zero emissions targets to limit global warming to below 1.5 °C. Improving energy efficiency is crucial for reducing carbon emissions, as energy consumption accounts for approximately three-quarters of these emissions. Cities, in particular, play a crucial role in implementing solutions for deep sustainable development [5]. However, the adoption of energy-efficient products has been slower than expected due to barriers such as a lack of investment, technology, and knowledge [6].

Rational energy use is an essential approach to sustainably managing resources, minimizing waste, and emissions. This involves implementing efficient technologies and adopting conservation practices across industrial, residential, and commercial sectors [7]. Education and awareness about efficient energy use are critical to overcoming existing barriers and promoting sustainable practices at the community level [8].

Energy consumption is intrinsically linked to economic and social development. However, inefficient use and depletion of fossil fuels pose significant environmental challenges. Adopting renewable energy sources and improving energy consumption efficiency are essential to ensuring long-term sustainability [9].

Buildings represent a significant opportunity for energy optimization, particularly in lighting systems. Proper planning and the use of efficient technologies can significantly reduce energy consumption and operational costs [10]. Different types of lamps, such as incandescent, halogen, fluorescent, and LED, vary in energy efficiency and lifespan, with LEDs being highly recommended for their efficiency and lower consumption [11, 12].

Therefore, the main objective of this research is to analyse energy efficiency and consumption within a building at the Technical University of Manabí. The study aims to identify strategies and technologies that optimize energy use, reduce costs, and minimize environmental impact, thereby contributing to sustainable development.

2 Methodology

The evaluation of energy efficiency in university buildings required an adapted methodology that considered various specific aspects of such constructions. Firstly, conducting a thorough literature review on energy consumption and efficiency in buildings was crucial. This step involved identifying parameters and methodologies used in previous studies to establish a robust theoretical foundation supporting the research.

Subsequently, specific objectives and the scope of the study were defined. This included setting clear goals for the technical–economic evaluation of energy efficiency and detailing the systems and specific aspects to assess, such as HVAC systems, lighting, and building envelopes.

The next critical step was data collection from the building. Detailed information, including plans, construction characteristics, and HVAC and lighting systems, was necessary. Historical energy consumption data was also collected to establish a baseline for accurate comparisons.

Energy analysis constituted a key stage in the methodology. This analysis was detailed, assessing the current energy consumption of the building and identifying areas for improvement, potential energy leaks, and inefficient systems. It was essential to consider both electrical and thermal consumption to gain a comprehensive understanding of the building's energy performance.

Next, specific measures to enhance the building's energy efficiency were identified. These measures included installing more efficient systems, improving the building envelope, and altering occupant usage habits. The energy to be generated by the photovoltaic system was evaluated with the Sunny Design tool to estimate the expected reduction in energy consumption and greenhouse gas emissions.

The analysis of implementation costs was another crucial aspect of the methodology. Determining the costs associated with implementing proposed measures, including equipment, materials, and labour, was necessary. Additionally, investment and operational costs over time were calculated, considering both initial costs and maintenance required to ensure the sustainability of implemented improvements.

Concurrently, an analysis of the economic and financial benefits derived from energy efficiency measures was conducted. This analysis included calculating savings in energy consumption and operational cost reductions. It was important to conduct a financial analysis that considered initial investment, return on investment (ROI), and other relevant financial indicators, thereby ensuring the economic viability of proposed measures.

The results are intended to include specific recommendations and an implementation plan supported by robust data and analysis. It is crucial that these recommendations be understandable to all stakeholders, including the university administration, to facilitate informed decision-making. Therefore, the technical and economic benefits of the proposed energy efficiency measures were emphasized. Subsequently, these measures were implemented, and a continuous monitoring system of the building's energy performance was established. This system utilizes the PQ-Box 150 A-Eberle Network Analyzer Measurement instrument, enabling assessment of energy quality and identification of consumption trends within the building. This monitoring facilitated adjustments to the measures based on observed results, ensuring the sustainability of the implemented improvements.

This methodology ensured a comprehensive and detailed approach to evaluate and improve energy efficiency in university buildings, aligning with recognized scientific and technical standards in the specialized literature (ISO 50001:2018 Standard, Energy Management Systems).

3 Case of Study

3.1 Weather Conditions of the City of Portoviejo, Manabí, Ecuador

Portoviejo is a city in the Coastal region, located in the Manabí Province of Ecuador, characterized by a dry tropical climate with maritime influences [13]. Portoviejo enjoys a considerable number of sunshine hours per year. On average, the city experiences approximately from 5 to 6 h of direct sunlight per day [14].

Temperatures in Portoviejo remain relatively stable throughout the year, with moderate seasonal variations. The average temperatures retrieved from Climate Consultant software are:

- January to March: 25–28 °C, with maximum temperatures reaching up to 32 °C.
- April to June: 24–27 °C, with highs near 30 °C.
- July to September: 23–26 °C, with peaks up to 29 °C.
- October to December: 24–27 °C, with maximums around 31 °C.

Nighttime temperatures rarely drop below 20 °C, influencing the need for both passive and active cooling systems.

Portoviejo experiences distinct climatic seasons, with a 6-month wet season (December to May) and a 6-month dry season (June to November) [15]. It is noteworthy that unlike other locations worldwide, Ecuador's wet season is generally hotter than its dry season [16].

Relative humidity in Portoviejo remains high throughout the year, ranging between 70 and 90%. This can increase thermal discomfort in the absence of adequate ventilation or air conditioning systems [17]. Predominant winds in the city are gentle, averaging speeds of 10 to 15 km/h [13]. During the dry season [15], winds may be more constant, enhancing natural ventilation and improving thermal comfort.

Direct solar radiation is another significant factor affecting thermal comfort, especially at noon, given Ecuador's location on the equator. Inadequate solar protection can lead to increased indoor temperatures. Portoviejo's climatic conditions necessitate careful consideration in architectural and urban design to ensure residents' thermal comfort. Factors such as proper ventilation, appropriate building materials, and the integration of green spaces are essential to mitigate the effects of heat and high humidity, creating a habitable and comfortable environment.

3.2 *Inventory of Electrical and Electronic Equipment in the Analysed Building*

The building, comprising four floors, has undergone a thorough inventory of active equipment, detailed in Tables 1, 2, 3, and 4.

The equipment catalogued includes lighting fixtures, air conditioning systems, printers, projectors, coffee makers, computers, microwaves, routers, emergency LED lamps, motion sensors, water dispensers, televisions, and the elevator. Technical specifications for each piece of equipment identified voltage, current, and corresponding power parameters.

Table 1 Inventory of electrical and electronic equipment on the ground floor

Item	Quantity	Voltage (V)	Electric current (A)	Power (KW)
Luminaires	122	82	1.78	0.170
Air conditioning	8	120	20	3.74
Printer	9	100	0.7	0.064
Projector	1	100	5.0	1.50
Computers	59	19.5	2.31	0.06
Coffee maker	1	120	20	3.27
Microwave	1	120	20	3.27

Table 2 Inventory of electrical and electronic equipment on the first floor

Item	Quantity	Voltage (V)	Electric current (A)	Power (KW)
Luminaires	88	82	1.78	0.170
Router	6	12	1.0	0.0165
LED Emergency lamps	4	120–220	0.02	0.006
Motion sensor	8	120	0.0045	0.0005
Water dispenser	4	120	5.5	0.915
Air conditioning	16	220	20	6.096
Printer	6	100	0.4	0.056
Projector	2	100	5.0	0.700
Computers	31	19.5	2.31	0.06

Table 3 Inventory of electrical and electronic equipment on the second floor

Item	Quantity	Voltage (V)	Electric current (A)	Power (KW)
Luminaires	130	82	1.78	0.170
TCL Air conditioning	8	208–230	20.2	4.260
CHIGO air conditioning	4	220	14.1	3.290
Projector	1	100	5.0	1.50
LED Emergency lamp	8	120–220	0.02	0.006

Table 4 Inventory of electrical and electronic equipment of the third floor

Item	Quantity	Voltage (V)	Electric current (A)	Power (KW)
Luminaires	133	82	1.78	0.170
Air conditioning	3	220	20	4.74
Projector	4	100	5.0	1.50
Computers (laptops)	3	19.5	2.31	0.05
LED Emergency lamp	5	120–220	0.02	0.006

4 Results and Discussion

4.1 Calculation of Energy Consumption in the Analysed Building

Luminaires. According to the data collected, the active power of a luminaire is 0.170 kW in a time period of 1 min. Thus, the energy consumption in 1 min is 0.002833kWh. Projecting this consumption in a month, considering that the lights are on for 10 h a day, and a national average rate of 9.2 US cents per kWh in Ecuador, the monthly cost for each lit luminaire is approximately $4.69 American dollars.

Air conditioning. The active power of an air conditioner is 3,74 kW in a period of 1 min, so the energy consumption in 1 min is 0.62333 kWh. Projecting this consumption in a month, considering that the air conditioners are on for 10 h a day during working hours, the monthly cost for each air conditioner is approximately $68.82 American dollars.

Printer. The active power of a printer is 0.064 kW in a period of 1 min. The energy consumption in 1 min is 0.001067 kWh. Projecting this consumption in a month, considering an average use of 2 h a day during working hours, the monthly cost for each printer is approximately $0.24 American dollars.

Projector. The active power of a projector is 1.50 kW in a period of 1 min.

The energy consumption in 1 min is 0.02500 kWh. Projecting this consumption in a month, considering an average use of 5 h a day during working hours, the monthly cost for each projector is approximately $13.80 American dollars.

Coffee maker and microwave. The active power of a coffee maker and a microwave is 3.27 kW in a period of 1 min. The energy consumption in 1 min is 0.05450 kWh. Projecting this consumption in a month, considering an average use of 1 h per day during working hours, the monthly cost for each coffee maker and microwave is approximately $6.02 American dollars.

Computer. The active power of a computer is 0.06 kW in a period of 1 min. The energy consumption in 1 min is 0.00100 kWh. Projecting this consumption in a month, considering an average use of 8 h a day during working hours, the monthly cost for each computer is approximately $0.88 dollars American dollars.

Router. The active power of a router is 0.0165 kW in a period of 1 min. The energy consumption in 1 min is 0.000275 kWh. Projecting this consumption in a month, considering a continuous use of 24 h a day, The monthly cost for each router is approximately $1.09 American dollars.

Motion sensor. The active power of a motion sensor is 0.0005 kW in a period of 1 min. The energy consumption in 1 min is 0.000008333 kWh. Projecting this consumption in a month, considering a continuous use of 24 h a day, the monthly cost for each motion sensor is approximately $0.03 American dollars.

Water dispenser. The active power of a water dispenser is 0.915 kW in a period of 1 min. The energy consumption in 1 min is 0.01525 kWh. Projecting this consumption in a month, considering an average use of 1 h per day during working hours, the monthly cost for each water dispenser is approximately $1.68 American dollars.

TVs. The active power of a television is 0.1 kW in a period of 1 min. The energy consumption in 1 min is 0.001666 kWh. Projecting this consumption in a month, considering an average use of 1 h per day during working hours, the monthly cost for each television is approximately $0.18 American dollars.

Elevator. The active power of an elevator is 5.8 kW in a period of 1 min. The energy consumption in 1 min is 0.09666 kWh. Projecting this consumption in a month, considering an average use of 24 h a day during working hours, the monthly cost per elevator is approximately $256.11 American dollars.

This analysis provides a detailed insight into the energy consumption and costs associated with different equipment in a four-story building in Ecuador, with a monthly consumption of 7,670.26 kWh and an expenditure of $6,164.762 (Table 5). Conducted at the Technical University of Manabí, the study's findings are likely applicable to other university buildings with similar characteristics. Thus, the results can help estimate energy consumption and costs in similar buildings, enabling more efficient energy resource planning and management. Additionally, the insights can guide the implementation of energy-saving measures, reduce operational costs, and support sustainable practices across the university.

Table 5 General consumption in Building #1 of the Faculty of Engineering and Applied Sciences of the Technical University of Manabí

Item	Quantity	Active Power (kw/1 min)	Monthly consumption (kWh/month)	Monthly cost per unit ($)	Total cost per month ($)
Luminaries	473	0.170	51	4.6920	2,219.316
Air conditioning (Ground floor)	8	3.740	748	68.8160	550.528
Air conditioning (1st floor)	16	6.096	1219.2	112.1664	1,794.662
Air conditioning (3rd floor)	3	4.740	948	87.2160	261.648
Printer	9	0.064	2.56	0.2355	2.120
Printer (1st floor)	6	0.056	2.24	0.2061	1.236
Projector	6	1.500	150	13.8000	82.800
Projector (1st floor)	2	0.700	70	6.4400	12.880
Coffee maker	1	3.270	65.4	6.0168	6.017
Microwave	1	3.270	65.4	6.0168	6.017
Computer	90	0.060	9.6	0.8832	79.488
Computers (3rd floor)	3	0.050	8	0.7360	2.208
Router	6	0.0165	11.88	1.0930	6.558
Motion sensor	8	0.0005	0.36	0.0331	0.265
Water dispenser	4	0.915	18.3	1.6836	6.734
TV	1	0.100	2	0.1840	0.184
Elevator	1	5.800	2784	256.1280	256.128
LED emergency	17	0.006	4.32	0.3974	6.756
TCL AA	8	4.260	852	78.3840	627.072
CHIGO AA	4	3.290	658	60.5360	242.144
Total			7,670.26		6,164.762

4.2 Proposals to Improve Energy Efficiency in the Analysed Building

Implementation of efficient lighting. Lighting represents one of the primary energy consumers in a building. Incandescent and fluorescent lamps, while common, have low energy efficiency compared to LED lamps. LED lamps consume significantly

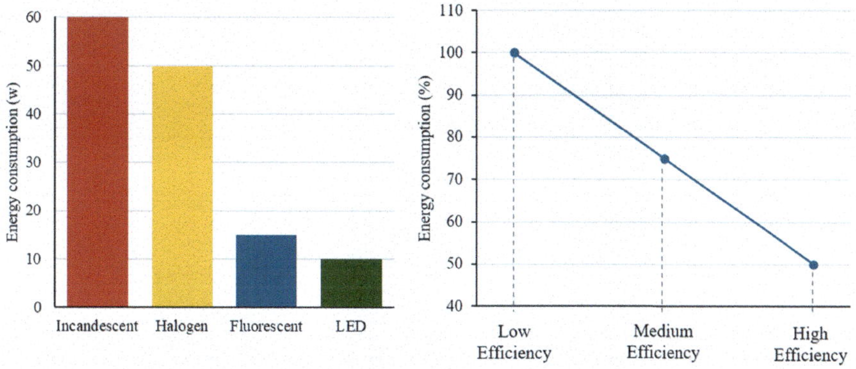

Fig. 1 Comparative energy consumption by type of lamp (left) and energy efficiency of HVAC systems. Reduction in energy consumption vs increase in efficiency in air conditioning systems (right)

less energy to produce the same amount of light and have a much longer lifespan, reducing the need for frequent replacements and thus maintenance costs.

Replacing incandescent and fluorescent lamps with LED lamps leads to significant benefits. LED lamps consume 50–80% less energy than traditional lamps and have a lifespan up to 25 times longer than incandescent ones. Additionally, LED lamps emit less heat, which reduces the load on HVAC systems and enhances thermal comfort in illuminated spaces. Figure 1 left illustrates the energy consumption of different lamp types.

Optimization of HVAC Systems. Air conditioning systems constitute a significant portion of the building's energy consumption. Upgrading these systems to more efficient models can significantly reduce energy consumption. Additionally, programmable thermostats and occupancy sensors enable automatic temperature adjustments based on occupancy schedules and presence of individuals, thereby avoiding unnecessary energy usage.

Installing high-efficiency air conditioning systems and using programmable thermostats and occupancy sensors to adjust temperatures as needed offer several benefits. These actions result in operational cost savings, improved thermal comfort within the building, and a reduction in energy consumption. Figure 1 right illustrates the reduction in energy consumption achieved by implementing these technologies.

Improved Thermal Insulation. Thermal insulation is crucial for maintaining a comfortable indoor temperature without relying excessively on heating or cooling systems. Effective insulation reduces heat loss in winter and heat gain in summer, achieved through the installation of insulating materials in walls and ceilings, and the use of double or triple-pane windows.

Applying thermal insulation to walls, ceilings, and windows, along with using double or triple-pane windows, offers numerous benefits. These actions reduce heat loss during winter and heat gain during summer, lower energy demand for heating

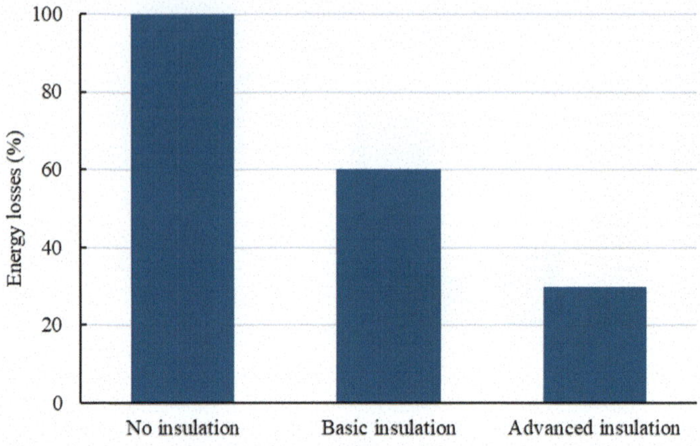

Fig. 2 Energy losses depending on insulation type

and cooling, and enhance indoor comfort. Figure 2 illustrates the reduction in energy losses with the use of advanced insulation.

Installation of Photovoltaic Solar Panels. The implementation of solar panels in the university building can significantly enhance energy efficiency, reduce operational costs, and decrease carbon footprint. Currently, there is a generation capacity of 3.4kWh on days with the highest solar radiation. The benefits of this system include a reduced dependence on the electrical grid, resulting in lower energy costs and a significant contribution to environmental sustainability.

Monitoring is conducted through the Geographic Information System for Sustainable Development of the Technical University of Manabí. Each device is located on the roof of Building 1 of the Faculty of Engineering and Applied Sciences, ensuring maximum efficiency of the solar panels and early detection of potential failures. In this regard, Fig. 3 illustrates the average energy generation per hour, and Fig. 4 displays the electrical parameters of the solar panels in Building 1 of the Faculty of Engineering and Applied Sciences.

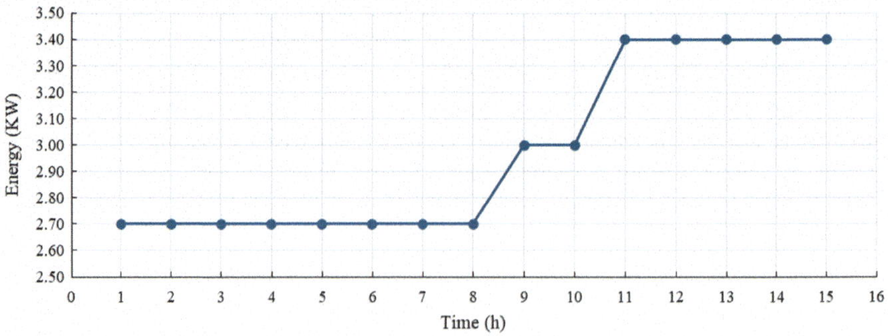

Fig. 3 Energy generated by the PV system vs the time elapsed in the day

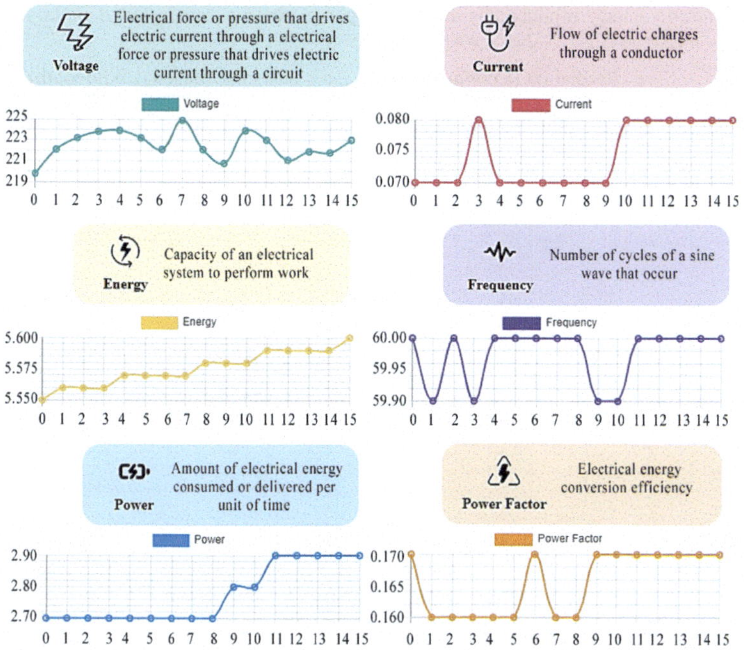

Fig. 4 Electrical parameters of the solar panels in building 1 of the faculty of engineering and applied sciences

Table 6 presents the calculated values of the monthly energy produced by the PV system, based on measurements of monthly irradiation retrieved from a Kipp&Zonen CMP11 pyranometer and generated through simulations using Sunny Design software.

As shown in Table 6, the energy produced by the PV panel system is minimal in November, with a monthly average of 8,613 kWh. Considering only the energy improvement from installing PV panels, even in the least favourable month, the energy produced by the PV panels would cover the building's energy needs since the building's monthly consumption is 7,670.26 kWh. In all other months, energy production exceeds this figure. The November output can be further improved by implementing additional measures such as upgrading insulation or changing the type of lighting used.

Table 6 Monthly energy produced by the 392 m^2 photovoltaic system installed on the terrace of the building analysed (Jinko Solar Tiger Pro JKM550M-72HL4-V Solar Panel)

Month	Irradiation (kWh/m^2/day)	Energy produced (kWh/month)
January	4.35	7,734
February	4.57	8,342
March	5.08	10,339
April	5.43	9,743
May	4.92	9,255
June	3.87	8,179
July	4.00	8,069
August	4.43	8,169
September	4.26	9,525
October	3.37	8,585
November	3.82	7,370
December	4.45	8,043

4.3 Energy Consumption and Economic Balance of the Improvement Proposals

It is imperative to critically analyse the proposed improvements for energy consumption and economic balance to understand their projected savings. While improved thermal insulation is recognized as a significant measure for enhancing energy efficiency, this proposal will not be analysed in the current study due to the inherent complexities in projecting its savings. Accurately quantifying the impact of improved thermal insulation involves numerous variables, including building design, local climate conditions, and existing insulation quality, which require extensive data and sophisticated modelling techniques. The variability and site-specific nature of these factors make it challenging to generalize the savings projections with the precision required for this analysis. Therefore, to maintain the scientific rigour and coherence of our study, we will focus on the other proposals where the projected savings can be more reliably estimated and validated (Implementation of efficient lighting, optimization of HVAC systems, and the installation of photovoltaic solar panels). This analysis will delve into the potential energy savings and cost reductions associated with these proposals, utilizing both empirical and statistical data to provide a comprehensive assessment of their effectiveness. By quantifying the impact of these improvements, we can better inform decision-making processes aimed at enhancing energy efficiency and economic viability.

LED Lamps versus Incandescent Lamps. LED lamps are significantly more energy-efficient than incandescent lamps, with a monthly consumption that is 46.50 kWh lower (Table 7). This reduced energy consumption translates into substantial cost savings, with the total monthly cost for LED lamps being $2074.83 less than that for incandescent lamps (Table 7). The cost–benefit ratio clearly favours LED lamps, making them a highly effective measure for reducing both energy use and costs.

Inverter Air Conditioners versus High-Efficiency Air Conditioners. Inverter air conditioners are more energy-efficient than their high-efficiency counterparts, saving 700.60 kWh per month (Table 8). This results in significant energy savings and a reduction in operational costs. The total monthly cost for inverter units is $161.36 less than for high-efficiency units (Table 8). The adoption of inverter air conditioners is a financially sensible and sustainable decision, as it reduces energy consumption and operational costs. Although the initial investment may be higher, the long-term savings justify the additional expense.

Photovoltaic Systems. The implementation of photovoltaic systems not only generates 103,353kWh/year but also reduces greenhouse gas emissions. These systems provide clean and unlimited electricity, allowing for the calculation of annual CO_2 emission reductions. The photovoltaic system installed at the Faculty of Engineering and Applied Sciences, with a capacity of 69.96 kWp, generates 103.35MWh annually, resulting in a reduction of 19.81 tons of CO_2 per year. This value is consistent with simulation results from Sunny Design, which estimate reductions of 21.38 tons and 34.70 tons of CO_2, respectively. This demonstrates the significant environmental benefits of adopting photovoltaic systems.

Therefore, the annual consumption of the building amounts to approximately 74000 kWh. The energy generated by photovoltaic panels will provide surplus energy that can be used in other university buildings, completely offsetting the electricity expenses of the analysed building. Together with the other proposed measures, this will create a positive impact on energy savings and environmental preservation, serving as a potential model for replication in other university buildings.

Table 7 Comparison of energy and economic consumption by types of lamps

Item	Quantity	Monthly consumption (kWh)	Unit value ($)	Total value per month ($)
Incandescent lamp	473	51.00	4.692	2,219.32
LED lamp	473	4.50	0.41	193.93

Table 8 Comparison of energy and economic consumption by types of air conditioners

Item	Quantity	Monthly consumption (KWH)	Unit value ($)	Total value per month ($)
High efficiency (ground)	8	748	68.816	550.53
High efficiency (1)	16	1,219.20	112.166	1794.66
High efficiency (3)	3	948	87.216	261.65
Inverter	27	2,214.60	203.79	2,445.48

5 Conclusions

Within Building 1 of the Faculty of Engineering and Applied Sciences, the electronic equipment that generates the highest economic demand is the air conditioning system, costing $2,606.84. In second place are the luminaires, with a cost of $2,219.32. In total, the budget allocated to the electricity service in this building amounts to $6,164.76. The implementation of efficient lighting systems and the use of renewable energy sources can yield significant economic savings and a notable reduction in energy consumption in the university building. By integrating LED lighting, the building has already achieved a reduction of 46.5 kWh. Additionally, the installation of photovoltaic systems is projected to generate approximately 29,000 kWh/year more than the building's energy needs. This surplus not only ensures a 100% reduction in electricity costs but also allows for the storage of excess energy. This stored energy can then be utilized in other buildings or for various other purposes, further enhancing the overall energy efficiency and sustainability of the university's infrastructure. Optimizing HVAC systems is identified as a key opportunity to enhance energy efficiency. This includes employing advanced technologies and appropriate maintenance practices that can decrease energy consumption and improve thermal comfort in the building. Investments in energy efficiency prove beneficial both economically and environmentally. Reduction in operating costs and contribution to long-term sustainability are among the key advantages highlighted in the study. The positive influence of incentive policies and financing is crucial for promoting energy efficiency in educational settings. These policies can facilitate the adoption of efficient technologies and sustainable practices, promoting greater awareness and commitment to energy efficiency.

References

1. Kosai S, Unesaki H (2024) Nuclear power, resilience, and energy security under a vulnerability-based approach. Clean Energy Syst 7:100107. https://doi.org/10.1016/j.cles.2024.100107
2. Cui Y, Xu Y, Wang Y et al (2024) Peer-to-peer energy trading with energy trading consistency in interconnected multi-energy microgrids: A multi-agent deep reinforcement learning approach. Int J Electr Power Energy Syst 156:109753. https://doi.org/10.1016/j.ijepes.2023.109753
3. Zheng J, Tu X, Xiang Y (2024) The impact of import expansion on energy efficiency: perspective of cities' geographical characteristics. J Clean Prod 437:140653. https://doi.org/10.1016/j.jclepro.2024.140653
4. Oyewole OJ, Al-Faryan MAS, Adekoya OB, Oliyide JA (2024) Energy efficiency, financial inclusion, and socio-economic outcomes: evidence across advanced, emerging, and developing countries. Energy 289:130062. https://doi.org/10.1016/j.energy.2023.130062
5. Zhang H, Zhou P, Sun X, Ni G (2024) Disparities in energy efficiency and its determinants in Chinese cities: from the perspective of heterogeneity. Energy 289:129959. https://doi.org/10.1016/j.energy.2023.129959
6. Lin B, Jia H (2024) Present-biased individuals and their underinvestment in household energy efficiency: evidence from first-tier Chinese cities. Energy Policy 185:113963. https://doi.org/10.1016/j.enpol.2023.113963

7. Bedoya Sánchez S, García Franco JF, Arango Lemoine C, Carvajal Quintero SX (2019) Análisis de programas de uso racional y eficiente de la energía en microrredes aisladas alrededor del mundo aplicación en Colombia. Grindda 1:149–157
8. Granados Guerrero HG, Duarte Rubio FE (2021) Sistema de control de luminarias y aires acondicionados y monitoreo de energía electrica y temperatura de aulas del edificio de posgrados de la universidad libre seccional cúcuta. Universidad Libre
9. Vaca Serrano JM, Kido Cruz A (2021) Estrategia de eficiencia en el consumo de energía eléctrica y mitigación en la estructura productiva de México. Contaduría y Administración 66(2):1
10. Hancevic P, Navajas F (2015) Consumo residencial de electricidad y eficiencia energética: Un enfoque de regresión cuantílica. Trimestre Econ 82:897–927. https://doi.org/10.20430/ete.v82i328.188
11. Kapoor G, Singhal M (2024) Impact of innovative thermal insulation materials in the building envelope on energy efficiency of residential buildings. Mater Today Proceed. https://doi.org/10.1016/j.matpr.2024.04.041
12. Yussuf RO, Asfour OS (2024) Applications of artificial intelligence for energy efficiency throughout the building lifecycle: an overview. Energy Build 305:113903. https://doi.org/10.1016/j.enbuild.2024.113903
13. Pourrut P (1983) Los climas del Ecuador: Fundamentos explicativos. CEDIG Documentos de Investigación
14. Delgado E (2020) Optimización energética de un bloque de viviendas de interés social en Ecuador. 140
15. Delgado D, Sadaoui M, Ludwig W, Méndez W (2022) Spatio-temporal assessment of rainfall erosivity in Ecuador based on RUSLE using satellite-based high frequency GPM-IMERG precipitation data. CATENA. https://doi.org/10.1016/j.catena.2022.106597
16. Delgado D, Sadaoui M, Pacheco H, et al. (2021) Interrelations between soil erosion conditioning factors in basins of Ecuador: contributions to the spatial model construction. In: International conference on water energy food and sustainability. Springer International Publishing, pp 892–903
17. Pourrut P (1994) L'eau en Equateur: principaux acquis en hydroclimatologie. ORSTOM éditions 30–42:1

Safety Conditions in Battery Rooms for Renewable Energy Systems: Dimensioning, Ventilation, and Classification of ATEX Areas

Jesús M. Ballesteros Álvarez, Álvaro Romero Barriuso, Blasa María Villena Escribano, Ángel Rodríguez Sáiz, and Cristina González Gaya

Abstract This chapter analyzes the safety conditions in battery rooms for renewable energy installations, focusing on sizing, ventilation, and classification according to the ATEX directive. For this purpose, the applicable European regulations are used as a reference to determine the classified areas according to fire and explosion risks, as well as the required safety distance and ventilation. An example application for a 24-V lead-acid battery is presented. The chapter also discusses safety measures for battery rooms that produce hydrogen and oxygen during the charging process, with reference to the technical reference specifications for determining the required hazard distance and ventilation openings. Graphical results are provided for the different battery types and voltages analyzed. Finally, it is concluded that the measures established by the new European application standard need to be adapted to comply with current legislation.

1 Introduction

The Thermal Installations Regulation in Buildings [1] mandates energy efficiency requirements, which can be met by using sustainable and efficient systems, including the integration of renewable energy sources in buildings. This requirement should apply not only to newly constructed buildings but also to the restoration and rehabilitation of existing historical structures. In such case, it is necessary to evaluate the feasibility of replacing fossil fuel-based energy systems with renewable energy

J. M. Ballesteros Álvarez · Á. Rodríguez Sáiz
Department of Architectural Constructions & Construction and Land Engineering, University of Burgos, Burgos, Spain

Á. Romero Barriuso (✉) · B. M. Villena Escribano · C. González Gaya
Department of Construction and Manufacturing Engineering, Universidad Nacional de Educación a Distancia (UNED), Madrid, Spain
e-mail: aromero@ind.uned.es

M. Torres González and C. Rubio Bellido (eds.), *Towards Low and Positive Energy Buildings*, Green Energy and Technology, https://doi.org/10.1007/978-3-031-70851-0_4

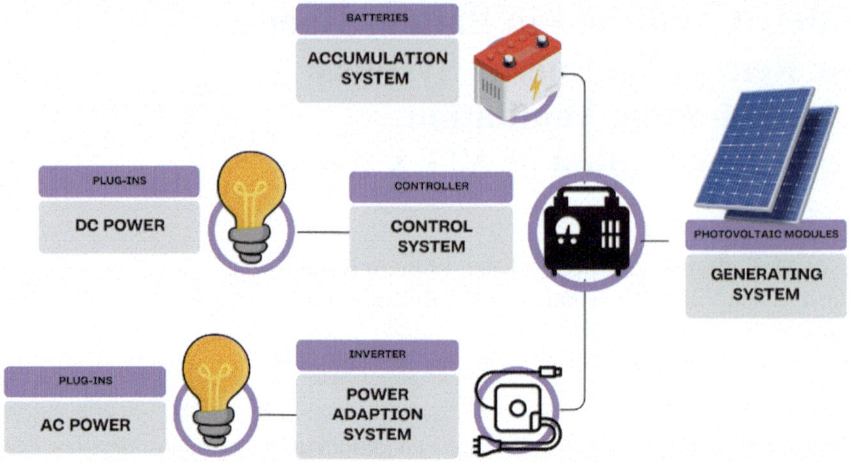

Fig. 1 General diagram of a photovoltaic installation with batteries. *Source* Own elaboration, 2023

installations. Failure to make this substitution requires a well-justified explanation [2].

One of the methods for generating renewable energy is through the use of photovoltaic panels that convert sunlight into electricity. To store this electricity, batteries are employed, which accumulate the electrical energy produced from the installation´s crystalline silicon or gallium arsenide cells on the collector plates [3]. This produces a clean and sustainable energy reserve, providing users with a certain degree of independence from non-renewable energy sources (Fig. 1).

Batteries used for electrical energy storage must be installed in enclosed enclosures that comply with the relevant regulations. This will ensure safety for personnel and equipment [4, 5]. These enclosures should feature unique construction characteristics, specialized electrical installations, and fire safety equipment [6]. Similarly, the enclosure must ensure optimal equipment operation under safe and environmentally friendly conditions by guaranteeing adequate environmental conditions [7].

The lead-acid battery is the most frequently used battery type in photovoltaic installations [8, 9]. It can be flooded or sealed with VRLA or gel valves [10]. Alternatively, lithium-ion batteries or flow batteries may be utilized, in which the electrolyte is stored in separate tanks. Lastly, sodium–nickel chloride batteries [11] can also be employed.

The implementation of this technology carries substantial handling risks in its handling risks due to a possibility of explosive gas concentration in the enclosures during operation. Preventive measures must be taken to minimize the risk [6, 12].

For instance, lead-acid and nickel batteries may produce emissions of hydrogen and oxygen gas while being charged due to the electrolysis of water created by the charging current. The release of these gases into the environment can form an explosive mixture if the hydrogen concentration in the air exceeds 4% by volume [13]. This situation necessitates implementing extra obligatory safety measures for workers exposed who encounter this hazard [14–16] and using of electrical equipment specifically designed for this type of risk [5, 17].

As previously indicated, the batteries should be kept covered and in protected areas, ideally in a separate installation room with limited access to maintenance personnel. In the event of utilizing open-type batteries, they can be placed in trays on an impermeable floor that is chemically resistant to the electrolyte. Ensure compliance with these guidelines for the safety storage of batteries.

To enable the dissipation of any electrostatic charges, any element at a distance of 1.25 m from the battery must have a resistance of less than 10 MΏ. During maintenance activities, personnel should wear antistatic footwear [13].

With the recent release of EN 60079-10-1:22 [5], which replaces EN 6079-10-1–10 [17], a challenge has emerged regarding the computation of the safety distance or hazardous distance. This is the specific vicinity in which machines that produce sparks and can trigger potential gaseous explosions should not be deployed. The safety requirements for battery installations are addressed in the EN 62485-2:19 standard [13]. However, this standard uses expressions from a repealed version for calculations. As a result, it is important to modify these expressions to align with the current version of the standard EN 60079-10-1-1:22 [5], which is presently in effect.

According to regulation [6], there exist three ATEX Zones with the potential risk of explosion resulting from gas accumulation of gases [4]. These zones are delineated as follows:

- Zone 0: An explosive atmosphere is expected to be present continuously, for long periods or frequently, with a probability of greater than 1000 hours per year.
- Zone 1 refers to an area where an explosive atmosphere may be present intermittently during regular operations. The likelihood of its occurrence falls between 10 and 1000 h per year.
- Zone 2: An area where the likelihood of an explosive atmosphere being present during normal operation is low, or if present, only for a brief duration. The probability of its existence is less than 10 h per year.

The purpose of this study is to identify the specific conditions that enclosures with installed batteries must meet in order to meet both constructive and safety requirements. Additionally, this study outlines the calculation process necessary for designing an enclosure with installed batteries within a classified zone, while establishing the required preventive measures to ensure the safety of the installation during operation.

2 Methodology

Fire and explosion risk areas should be classified based on objective reference factors, including hazardous distance, enclosure ventilation, necessary ventilation airflow, and potential gas escape in areas close to the batteries where dilution is not assured.

The ventilation airflow rate should also be determined, by considering the number and type of installed batteries and the potential for explosive gas concentration [13]. It is important to use the equations (1) and (2):

$$Q = 0.05 \cdot n \cdot I_{\text{gas}} \cdot C_{rt} \cdot 10^{-3} \tag{1}$$

$$Q = v \cdot q \cdot s \cdot n \cdot I_{\text{gas}} \cdot C_{rt} \cdot 10^{-3} \tag{2}$$

where:

$Q =$ Ventilation airflow (m^3/h)

$v =$ Necessary dilution of hydrogen $v = \frac{100\% - 4\%}{4\%} = 24$.

$q =$ Generated hydrogen at 0 °C (m^3/Ah) $q = 0.42 \cdot 10^{-3}$.

Remark: for calculations at 25 °C, the value of q at 0 °C shall be multiplied by factor 1,095.

$s =$ general safety factor $s = 5$.

$n =$ Is de number of cells

$I_{\text{gas}} =$ Is the current producing gas-rated capacity (mA/Ah)

I_{float} For the float charge current (mA/Ah)

I_{boost} The boost charge current (mA/Ah)

$C_{\text{rt}} =$ Is the C10 capacity for lead-acid cells (Ah), Uf = 1.80 V/cell at 20 °C or C5 capacity for NiCd cells (Ah), Uf = 1.00 V/cell at 20 °C.

The exhaust characteristic value serves as an indicator for hazardous site classification. To determine this value, the exhaust source, referring to the flammable substance emitted during the process, must be characterized based on factors related to enclosure ventilation and gas flammability or explosiveness. The expression utilized for this purpose is Eq. 3 [4]:

$$\text{Exhaust Characteristic}\left(m^3/s\right) = \frac{Q}{f \cdot 3600 \cdot \rho_g} \cdot \frac{\text{LIE}_m}{L_{II}} \tag{3}$$

where:

$Q =$ Ventilation airflow (m^3/h)

$f =$ Measurement of the degree of mixing between exhaust and ventilation air

$\rho_g =$ Gas density (Kg/m^3)

$L_{\text{II}} =$ Lower flammability limit (vol/vol)

$\text{LIE}_M =$ Lower mass explosive limit (Kg/m^3).

Next, determine the safety distance (Eq. 4), which is the area where devices emitting sparks or with incandescent properties must not be used, while accounting for the hemispherical gas dispersion [13].

$$d = 28.8 \cdot \sqrt[3]{I_{\text{gas}}} \cdot \sqrt[3]{C_g} \text{ mm} \tag{4}$$

where:

$d =$ Dangerous distance (mm)

Igas = Gas-producing stream (mA/Ah)

Crt = Allocated capacity (Ah).

When the gases from a monobloc battery containing n cells are discharged through a single opening, the safety distance calculation should be based on the value obtained from this opening (Eq. 5):

$$C_{rt} = C_{rt} \cdot n \tag{5}$$

where:

n = Number of cells.

The airflow acquired defines a hypothetical volume, assuming once again that the gas dispersion is hemispherical. Its volume is determined by the following expression (Eq. 6):

$$V_z = \frac{2}{3}\pi \cdot d^3 \tag{6}$$

where:

d = Safety distance from the gas emission source (mm).

With both expressions, it is obtained Eq. 7:

$$d^3 = \frac{3}{2 \cdot \pi} \cdot 0.05 \cdot n \cdot I_{\text{gas}} \cdot C_{rt} \cdot 10^6 \tag{7}$$

Finally, the necessary ventilation openings in the battery room are calculated by utilizing the following Equation (Eq. 8) [13].

$$A = 28 \cdot Q \tag{8}$$

where:

Q = Ventilation airflow (m^3/s)

A = Free area of air inlet and outlet openings (cm^2).

The entire calculation process was conducted by considering the Regulations for the Installation of Batteries in Confined Spaces in Buildings, alongside the regulations that establish Areas with Explosive Atmospheres (EN IEC 62,485-2, 2019; EN 60,079-10-1, 2022) and the previous regulations that have already been repealed [17–19].

A comparison is drawn between the minimum volumetric fresh airflow rate (Eq. 9) stipulated in the previous standard and the proposed ventilation airflow rate (Eq. 10) in the updated version (EN IEC 62485-2, 2019).

$$f \cdot \left({}^{\mathrm{d}V}/_{\mathrm{d}t}\right)_{\min} = \frac{Q}{3600} \tag{9}$$

$$f \cdot \left({}^{\mathrm{d}V}/_{\mathrm{d}t}\right)_{\min} = \frac{v \cdot q \cdot s \cdot n \cdot I_{\text{gas}} \cdot C_{rt} \cdot 10^{-3}}{3600} \tag{10}$$

where:

Q = Ventilation airflow (m^3/h)

f = Measurement of the degree of mixing between exhaust and ventilation air

$(dV/dt)_{min}$ = Minimum fresh air volume flow rate (m^3/s)

v = Hydrogen solution required $v = \frac{100\%-4\%}{4\%} = 24$.

q = Hydrogen generated at 0 °C (m3/Ah) $q = 0.42 \cdot 10^{-3}$.

For the calculation at 25 °C, this value would have to be multiplied by 1.095

s = Safety factor s =5

n = Number of battery elements

I_{gas} = Current producing gas per rated capacity Ah (mA/Ah)

C_{rt} = C10 capacity of lead elements, Uf = 1,80V/element at 20 °C o the C5 capacity of NiCd elements, Uf 1.00V/element at 20 °C (Ah).

First of all, it is needed to characterize the exhaust, and to do so we proceed as follows (Eq. 11):

$$\text{Exhaust characteristic} = \frac{W_g}{\rho_g \cdot k \cdot L_{II}} \tag{11}$$

The minimum theoretical ventilation flow rate needed hydrogen gas, a flammable substance, to a concentration below the lower explosion limit (LEL), can be determined by using the Eq. 12:

$$\left({}^{dV}/_{dt}\right)_{min} = \frac{\left({}^{dG}/_{dt}\right)_{max}}{k \cdot \mathrm{LIE}_m} \cdot \frac{T}{293} \tag{12}$$

where:

Q = Ventilation airflow (m^3/h)

$(dV/dt)_{min}$ = Minimum fresh air volume flow rate (m^3/s)

$(dG/dt)_{max}$ = Maximum escape rate (Kg/s)

k =0.25 Safety factor applied to the LIEM, for emissions produced by the batteries

LIE_M = Lower mass explosive limit (Kg/m^3)

T = Ambient temperature, in degrees Kelvin (K)

W_g = Gas mass escape rate (Kg/s)

L_{II} = Lower flammability limit (vol/vol).

For calculating the maximum escape ratio, Eq. 13 is utilized (Eq. 13):

$$\left({}^{dG}/_{dt}\right)_{max} = w_g \tag{13}$$

Therefore, a connection is established between the phrasing found in the prior regulation (Eq. 14) [17] and the present regulation (Eq. 15) [5].

$$w_g = \left({}^{dV}/_{dt}\right)_{min} \cdot k \cdot \mathrm{LIE}_m \tag{14}$$

$$w_g = \frac{Q}{f \cdot 3600} \cdot k \cdot \text{LIE}_m \tag{15}$$

Consequently, the exhaust can be characterized by the airflow rate specified in the ventilation guidelines presented in the EN 62485-2:19 standard (Eqs. 16 and 17).

$$\text{Exhaust characteristic} = \frac{Q \cdot k \cdot \text{LIE}_m}{f \cdot 3600 \cdot \rho_g \cdot k \cdot L_{II}} \tag{16}$$

$$\text{Exhaust characteristic} = \frac{Q}{f \cdot 3600 \cdot \rho_g} \cdot \frac{\text{LIE}_m}{L_{II}} \tag{17}$$

3 Case Study

To enhance comprehension of the calculation procedure, we will analyze a frequently employed 24-V lead-acid battery with open-type elements and the subsequent load capacity attributes: C100: 900 Ah / C10: 686 Ah / C5: 590 Ah / C3: 510 Ah/ C1: 353 Ah.

In the case of lead-acid batteries, the discharge regime considered is C10, while for nickel–cadmium batteries, it is C5. For the particular battery selected, the C10 discharge regime results in its complete discharge in 10 h with a current of 68.6A.

The ventilation airflow rates for a site containing batteries can be calculated by utilizing the current value *I*gas, charge capacity C10, and the values of U_{float} and Fast $_{\text{Uptime}}$ for 24 V lead-acid batteries with a U_{float} of 2.23 and a Fast $_{\text{Uptime}}$ of 2.40 v/element. The following formula (Eq. 18) should be used:

$$Q = 1,7^{m^3}/_h \tag{18}$$

If the manufacturer fails to provide values, please consult Table 1 of the reference standard for the appropriate values for the various types of tamping troughs covered by the standard.

The obtained ventilation flow rate permits classification of hazardous areas by their risk of fire or explosion using the exhaust value (Eq. 19).

$$\text{Exhaust characteristic} = \frac{1.7}{1.25 \cdot 3600 \cdot 0.089} \cdot \frac{0.0033}{0.04} = 3.5 \cdot 10^{-4 m^3}/_s \tag{19}$$

During battery charging, primary leaks may lead the development of an atmosphere around the battery caps. For installations designed for simultaneous charging of multiple batteries with multiple exhausts, it is imperative to follow the criterion established for multiple exhausts [17].

Table 1 Presents the characteristic values necessary to determine the airflow rate for ventilation at a site with coils. *Source* [13]

Parameter	Lead-acid batteries open type elements	Lead-acid batteries VRLA valve-regulated cells	NiCd Batteries open-type elements
f_{gas}	1	0.2	1
Fs	5	5	5
U_{float} (v/element)	2.23	2.27	1.40
I_{float} (mA/Ah)	1	1	1
I_{gas} (mA/Ah)	5	1	5
U_{fast} (v/element)	2.40	2.40	1.55
I_{fast} (mA/Ah)	4	8	10
I_{gas} (mA/Ah)	20	8	50

The standard mandates an airflow rate of 0.1 m/s when the primary exhaust rate is below 0.001 m^3/s. This requirement indicates a high degree of dilution, resulting in a flammable volume of less than 0.1 m^3. Subsequently, for a primary exhaust system with high dilution and natural ventilation, a Zone 1 (E.D.) + Zone 2 classification is achieved. Whereas, for artificial ventilation, the result leads to a non-hazardous Zone 1 (E.D.).

The EN 202007: IN [16] standard is used to analyze multiple exhausts (Eq. 20). It states that only the exhausts obtained by using the following expression should be considered as simultaneous if the total number of primary exhaust sources exceeds fifty.

$$m = 12 + 0.2 \cdot (n - 51) \tag{20}$$

where:

m = Number of primary leaks to consider

n = Total number of primary leaks.

Therefore, to achieve average dilution in a bathhouse with an air velocity of 0.1 m/s and an exhaust velocity of 0.007 m^3/s, at least 20 primary exhausts need to be installed.

This would be equivalent to the number of batteries possessing comparable characteristics (Eq. 21).

$$20 = 12 + 0.2 \cdot (n - 51), \tag{21}$$

where n=91

Without proper ventilation in the room, the air velocity inside would be roughly at 0.05 m/s, a figure that is generally applicable to small environments. Under these circumstances, with 24 coils, the dilution can be considered medium, as exhaust would exceed 0.003 m^3/s resulting in the equivalent of 8 primary exhausts. If only natural ventilation is available, the situation would result in a Zone 1 + Zone 2 classification, while artificial ventilation would lead to Zone 1.

To determine the extent of the hazardous area, the standard includes graph for calculation that distinguishes among three exhaust behaviors: high-velocity free jet exhaust, low-velocity diffusion jet exhaust, or gases and vapors that spread along horizontal surfaces (heavy gases). The exhaust in this installation is a jet or obstructed jet exhaust, i.e., diffusive, and the hazardous distance obtained graphically is 1 meter [5].

The hazardous area for hydrogen has a conical geometry that flattens at the top due to its low-pressure nature, allowing it to rapidly dilute because of its low relative density. Furthermore, the gas is emitted from each of the battery valves, thus indicating that the danger zone corresponds to a hemisphere situated above the radio *dz* battery.

To establish a safe distance, the standard suggests using the following Equation (Eq. 22):

$$d = 28.8 \cdot \sqrt[3]{I_{\text{gas}}} \cdot \sqrt[3]{C_g}\text{mm} \tag{22}$$

where:

d = Dangerous distance (mm)

I_{gas} = Current producing gas (mA/Ah)

C_{rt} = Allocated capacity (Ah).

When gases are released from a monoblock battery containing *n* cells through a single opening, the value utilized to compute the safety distance from such opening must be determined (Eq. 23).

$$C_{rt} = C_{rt} \cdot n \tag{23}$$

where:

n = Number of cells

The airflow rate generated determines a hypothetical volume, assuming a hemispherical gas dispersion. To calculate the volume, utilize the following expression (Eq. 24):

$$V_z = \frac{2}{3}\pi \cdot d^3 \tag{24}$$

where:

d = Safety distance from the gas emission source (mm)

Joining the two expressions, it is obtained Eq. 25:

$$d^3 = \frac{3}{2 \cdot \pi} \cdot 0.05 \cdot n \cdot I_{\text{gas}} \cdot C_{rt} \cdot 10^6 \tag{25}$$

Applying the given formula to the battery selected as a model yields the following findings (Eqs. 26 and 27):

$$d^3 = \frac{3}{2 \cdot \pi} \cdot 0.05 \cdot 10 \cdot 5 \cdot 686 \cdot 10^6 \tag{26}$$

$$d = 935 \text{ mm} \; = \; 0.93 \text{ meters} \tag{27}$$

The plant room can be ventilated naturally or through forced ventilation. Natural ventilation is the preferred option and requires an air inlet and outlet with a determined minimum free area opening. This opening can be calculated using Eq. 28:

$$A = 28 \cdot Q \tag{28}$$

where:

Q = Ventilation airflow (m^3/s)

A = Free area of air inlet and outlet openings (cm^2).

The air intake and exhaust will be positioned to optimize airflow and will be either on opposing walls or have a minimum separation distance of two meters if they are situated on the same wall.

Forced ventilation must be included in the room when it is impossible to achieve the necessary airflow through natural ventilation. The charger must be interlocked with the ventilation system, and the air from the battery room must be directed outside the building to ensure safety and proper ventilation. Additionally, the extraction system must possess the ATEX marking designated for hydrogen.

Openings with specific dimensions are required in this scenario (Eq. 29):

$$A = 28 \cdot Q \rightarrow \mathrm{A} = 47.6 \text{ cm}^2 \tag{29}$$

4 Results

The charge capacity values utilized in the application example are prevalent among batteries regularly deployed in photovoltaic installations.

To enhance comprehension of the hazardous distances and the ventilation opening dimensions, we present graphical results for batteries with varying charge capacities and potential differences, segregated by type. Figure 2 illustrates the hazardous distances of various battery models that pose a risk of hydrogen emission during charging.

The data indicates a consistent trend among the three battery types, with a more pronounced increase in the hazardous distance between lower voltages. Batteries with exposed components create a greater hazard distance. Lead-acid batteries necessitate a larger safety distance.

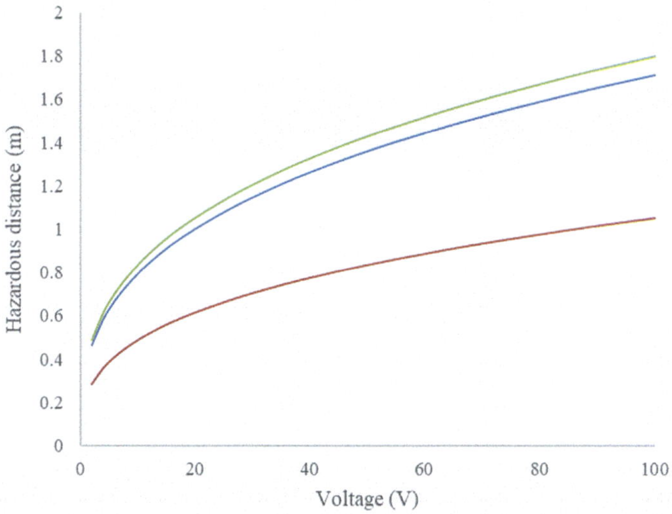

Fig. 2 Hazardous distance is depicted as a function of voltage in the figure. Lead-acid batteries with open cells are represented by the green curve, while nickel–cadmium batteries with open cells are represented by the blue curve. The red curve, on the other hand, represents lead-acid batteries with VRLA valves

The results for the required free area of ventilation for different battery models analyzed in this chapter are presented in Fig. 3. It shows a linear trend indicating a greater need for ventilation in batteries with open elements, with nickel–cadmium batteries being particularly notable.

5 Conclusions

Accumulators and batteries utilizing aqueous electrolytes generate hydrogen and oxygen as a result of the overcharge current passing through during the charging process. This requires implementing the safety precautions outlined in the EN 62485-2:2019 standard [13] to prevent direct and indirect electrical contact, including automatic disconnection of the power supply and measures to prevent short circuits. Furthermore, at times, the emission of hydrogen into the atmosphere can create a classified zone for fire and explosion. In this case, additionally, the requirements established in Royal Decree 144/2016 [16] would have to be implemented, among which the specific marking for electrical and non-electrical equipment stands out. For this particular case, it would be Ex II 2G IIC T1 Gb, the appropriate marking for enclosures that are permitted to protect electrical equipment within a hydrogen-hazardous environment are: Ib, d, e, m, o, p, and q, while the possible enclosures for non-electrical equipment are d, c, b, and k.

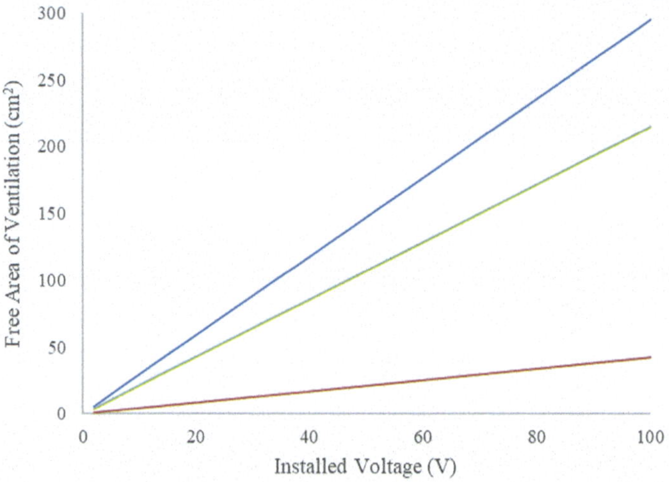

Fig. 3 Free area required for ventilation as a function of voltage. The green curve represents lead-acid batteries with open elements. The blue curve represents nickel–cadmium batteries with open cells. The red curve represents lead-acid batteries with VRLA (Valve Regulated Lead-Acid) technology

The existence of the classified area necessitates consideration of the obligations specified in Royal Decree 681/03, dated June 12, with respect to worker access to hazardous areas, provision of preventive resources, authorization, specialized training in ATEX, and the availability of appropriate protective gear, such as anti-static footwear. Furthermore, it is crucial to emphasize the requirement for drafting an explosion protection document [14, 16].

According to the aforementioned, an explosive atmosphere can arise from the generated gases when the concentration of hydrogen in the air surpasses 4%. To ensure safety, it is crucial to determine the hazardous distance. EN 62485-2:19 [13] bases its calculation on an outdated version of EN 60079-10-1:10 [17]. This version differs significantly from 2016 version cited in Royal Decree 681/2003 and the most recently published version in 2022, which transitioned from numerical to graphical resolution for classified areas.

Therefore, based on the previous discussion, we propose an alternative approach for applying standard EN 62485-2:19 [13] that is adjusted to the latest version of EN 60079-10-1-1:22 [5]. This proposal aligns with the provisions of Royal Decree 681/2003 [14].

Additionally, we include the resolution of a practical example that facilitates its implementation. The resulting graphical outcome illustrates need to modify the measures established by the EN 62485-2:19 standard [13] for all the three battery types examined.

References

1. RITE-217. Real Decreto 1027/2007, de 20 de julio, por el que se aprueba el Reglamento de Instalaciones Térmicas en los Edificios (2013). Ministerio de la Presidencia. Gobierno de España
2. Márquez JM, Moreno S, Acha CA, Jubera J (2014) Progress towards the implementation of a low-carbon economy in the states of the European Union. Rev Eur de Inv Arquitectura 5:80–94
3. Salazar A, Pichardo A, Pichardo U (2016) La energía solar, una alternativa para la generación de energía renovable. Rev Inv Desarrollo 2(5):11–20
4. NTP-369 (1995) Atmósferas potencialmente explosivas: instalaciones eléctricas. Instituto Nacional de Seguridad y Salud en el Trabajo (INSST). Ministerio de Trabajo y Economía Social
5. EN 60079–10–1:2022 Explosive atmospheres—Part 10–1: Classification of areas—Explosive gas atmospheres (2022) European Committee for Standardization: Brussels, Belgium
6. Directiva 2014/34/UE del Parlamento Europeo y del Consejo (2011) de 26 de febrero de 2014, sobre la armonización de las legislaciones de los Estados miembros en materia de aparatos y sistemas de protección para uso en atmósferas potencialmente explosivas (ATEX 100). Parlamento Europeo y Consejo de la Unión Europea. Unión Europea
7. NTP-617: Locales de carga de baterías de acumuladores eléctricos de plomo-ácido sulfúrico. Instituto Nacional de Seguridad y Salud en el Trabajo (INSST) (2003) Ministerio de Trabajo y Economía Social
8. Pávlov D (2011) Lead-acid batteries: science and technology. Elsevier
9. Alba CA, Muñoz MG (2017) Estudio y fabricación de una batería ácido plomo. Temas de Ciencia y Tecnol 21(6):23–28
10. Fullea J (2003) Acumuladores electroquímicos. McGraw Hill
11. Santillán A (2016) Estudio de la incorporación de baterías en sistemas fotovoltaicos. Trabajo Fin de Máster. Universidad de Barcelona, Barcelona, España
12. Sánchez A (2007) Requerimientos debido a la existencia de Atmósferas Explosivas. Dyna 82(3):41–44
13. EN IEC 62485–2:2019 Safety requirements for secondary batteries and battery installations—Part 2: Stationary batteries (2019) European Committee for Standardization: Brussels, Belgium
14. Real Decreto 681/2003 de 12 junio, sobre la protección de la salud y la seguridad de los trabajares expuestos a los riesgos derivados de atmósferas explosivas en el lugar de trabajo. Trasposición de la Directiva 99/92/CE (ATEX-137). Ministerio de la Presidencia. Gobierno de España.
15. NTP-876: Evaluación de los riesgos específicos derivados de las atmósferas explosivas (ATEX) (2010) Instituto Nacional de Seguridad y Salud en el Trabajo (INSST). Minist de Trabajo y Econ Soc
16. Real Decreto 144/2016, de 8 de abril, por el que se establecen los requisitos esenciales de salud y seguridad exigibles a los aparatos y sistemas de protección para su uso en atmósferas potencialmente explosivas y por el que se modifica el Real Decreto 455/2012, de 5 de marzo, por el que se establecen las medidas destinadas a reducir la cantidad de vapores de gasolina emitidos a la atmósfera durante el repostaje de los vehículos de motor en las estaciones de servicio. Ministerio de Energía, Industria y Turismo. Gobierno de España
17. UNE 202007:2006 IN Application guide of UNE-EN 60079–10. Electrical apparatus for explosive gas atmospheres (2006). Classification of hazardous areas. Asociación Española de Normalización y Certificación (AENOR). Madrid
18. EN 60079–10–1:2010 Explosive atmospheres—Part 10–1: Classification of areas—Explosive gas atmospheres. European Committee for Standardization: Brussels (2022) Belgium, (Derogada)
19. Ballesteros Álvarez JM, Romero Barriuso A, Villena Escribano BM, Rodríguez Sáiz A, González-Gaya C (2024) Investigating the effectiveness of a new indoor ventilation model in reducing the spread of disease: a case of sports centres amid the COVID-19 pandemic. Heliyon 10(6):2405–8440. https://doi.org/10.1016/j.heliyon.2024.e27877

Energy Performance Assessment and Thermal Comfort

Mapping the Energy Performance of Existing Residential Building Stock: The Case of Seville

María Fernanda Guerrero Rivera, Antonio Jesús Aguilar, Joaquín Aguilar Camacho, and María Luisa de la Hoz Torres

Abstract Residential buildings account for a significant share of the total energy consumption in the European Union (EU). With the European Commission estimating that 75% of Europe's building stock is energy inefficient, renovation of existing buildings is crucial to help reduce global energy consumption and CO_2 emissions. This study provides insight into the energy performance of the residential building stock in the most populated city of Andalusia, Seville. The data contained in 45,920 energy performance certificates (EPC) of dwellings were processed and analysed according to the type of dwellings (single-family dwellings and individual dwellings in multi-family blocks), construction period, and their district location. The results obtained revealed that most of the dwellings have a poor energy performance rating (with approximately 66% classified under label E). The districts with the highest percentages of dwellings with the worst-energy performance (EPC "G" and "F") are D4 Cerro-Amate district (36%) and D2 Macarena district (34%). The bottom-up approach used in this study provides useful information to assess the energy performance of Seville's residential buildings on the urban scale. These findings can help define renovation strategies and public energy policies targeting districts having buildings with poor energy performance.

M. F. Guerrero Rivera · A. J. Aguilar
University of Seville, Seville, Spain

J. Aguilar Camacho
Department of Graphic Engineering, University of Seville, Seville, Spain

M. L. de la Hoz Torres
Department of Building Construction II, University of Seville, Seville, Spain

A. J. Aguilar · M. L. de la Hoz Torres (✉)
Department of Signal Theory and Communications, Rey Juan Carlos University, Fuenlabrada, Spain
e-mail: maria.delahoz@urjc.es

M. Torres González and C. Rubio Bellido (eds.), *Towards Low and Positive Energy Buildings*, Green Energy and Technology, https://doi.org/10.1007/978-3-031-70851-0_5

1 Introduction

The building sector in the EU is responsible for 40% of the energy consumption and 36% of greenhouse gas emissions, arising from the different phases of the building's life cycle (i.e. construction, usage, renovation, and demolition) [1]. Concerning the total energy consumption by this sector, the upward trend in the consumption of European households from 1990 (26,45%) to 2022 (29,58%) is noteworthy. Although slightly lower values were found in Spanish households, an increasing trend can be identified there too (16,83% and 20,38% in 1990 and 2021, respectively) [2].

The energy use behaviour of households, along with citizens' awareness in this field, plays a crucial role in maintaining the well-being of homes [3]. A significant part of the energy consumed by households is used for air conditioning and heating. The International Energy Agency pointed out that the use of heating, ventilation, and air conditioning (HVAC) and domestic hot water (DHW) systems accounts for almost half of global energy use in buildings [4]. Since people spend a large part of their life indoors (e.g. at home, the workplace, at school, or other public buildings), ventilation [5], heating, and cooling indoor spaces are essential to achieve proper indoor environmental conditions, to ensure occupants' comfort, satisfaction, and health [6–8]. However, the use of HVAC systems is also a significant driver of global energy demand [4].

Directive 2010/31/EU of the European Parliament and of the Council of 19 May 2010 on the Energy Performance of Buildings [9], and Directive 2012/27/EU of the European Parliament and of the Council of 25 October 2012 on energy efficiency [10] stated the framework of the energy performance of buildings. Recently, the EU's target regarding energy efficiency was raised in the revised Energy Efficiency Directive [11]. Directive EU/2023/1791 [11] raised the energy savings obligations by 2028 and prioritises energy efficiency improvements for vulnerable customers, individuals affected by energy poverty and those living in social housing. This directive [11], together with the Energy Performance of Buildings Directive EU/2024/1275 [12], establishes the current legislative framework on the energy performance of buildings and promotes policies aimed at [12]: (1) Achieve a fully decarbonized building stock by 2050, (2) Help create a stable environment for investment decisions, and (3) Enable consumers and businesses to make more informed decisions to save money and energy.

These goals are in line with the European Green Deal [14] and the United Nations Agenda 2030 [15]. The Sustainable Development Goals call for action and urge nations to redouble their efforts to ensure universal access to affordable and modern energy services (SDG 7—Affordable and clean energy [16]).

These ambitious targets, which will help improve people's health and quality of life, also pose a challenge to European Member States, where 85% of the buildings were built before 2000 [13]. In Spain, 55% of the buildings were built before 1950 (roughly 21% are aged more than 50 years) and almost 58% were built before the first law that introduced minimum energy efficiency criteria in Spain [17]. Table 1 shows the evolution of Spanish regulations in the field of energy performance of buildings. The evolution of the regulations has been progressive: it was not until 1969 that specific measures were established, which have gradually increased the energy efficiency requirements up to the current level. This evolution has resulted in the definition and application of new thermal parameters and stricter requirements. Thus, the entry into force of the NBE-CT (1979) and the CTE (2006) are key dates and therefore will be taken into account in the assessment of the results obtained in this study. The EPC was required by the European Union through the 2010/31/EU Energy Performance of Buildings Directive [9].

In Spain, the basic procedure for EPC of new and existing buildings was approved by Royal Decrees 47/2007 and Decree 235/2013, respectively. The EPC is a label based on non-renewable energy consumption (EP_{nr}) and CO_2 emissions. The labels range from A (the most efficient building) to G (the most inefficient). The transition values between the different categories depend on the climate zone and the type of building.

The EPC label facilitates the comparison of different buildings and provides useful information about the building stock. It can also be used to define minimum energy performance standards (MEPS) as mentioned in the proposal introduced by the European Commission to the revision of the Energy Performance of Buildings Directive [25]. For example, France implemented a MEPS and established that the "F" and "G" dwellings must be EPC "B" by 2030, followed by "D" and "E" by 2040 [26].

In addition, EPCs can be used to identify energy-inefficient buildings at the urban scale. Geographic information system technology is very useful in this context, for example, to illustrate the regional mapping of energy consumption in buildings [27], to develop geographic indicators of energy consumption [28] or to identify buildings with poor energy performance [29]. In these buildings, high energy consumption, coupled with high energy costs and low household income, directly impacts energy

Table 1 Summary of the evolution of laws Energy Performance of Buildings in Spain

Year	Document	Main contents in terms of energy efficiency of buildings
1969	Law 69 [18]	First law that set requirements
1979	NBE-CT 79 [19]	Envelope quality requirements
2006	Technical building code CTE DB-HE 2006 [20]	Quality of the envelope (reference building demand) Renewable energies
2007	Royal decree 47/2007 [21]	Introduction of EPC for new buildings
2013	Royal decree 235/2013 [22]	Introduction of EPC for existing buildings
2013	Modification of technical building code CTE DB-HE 2013 [23]	Non-renewable primary energy consumption Heating and cooling demand Quality of the envelope Renewable energies
2019	Modification of technical building code CTE DB-HE 2019 [24]	Non-renewable primary energy consumption Total primary energy consumption Additional conditions (minimum building quality and minimum renewable energy contribution)

Table 2 Comparison of indicators obtained from Andalusian and national level based on 2021 data [35]

Indicators	% of household	
	Spain	Andalusia
Late payment of housing supplies bills	9.5	12.9
Inability to maintain the home at an adequate temperature during winter season	14.3	18.0
Twice the median (2 M)	16.4	24.5

poverty [30]. This issue is especially relevant in Spain [31, 32]. The EU statistics on income and living conditions (EU-SILC) revealed that 11,1% of Spanish households were unable to keep their homes adequately warm in 2014, a percentage that increased to 20.8% in 2023 [33]. In Spain, Andalusia is one of the regions with the highest levels of energy poverty [34]. A comparison of different indicators obtained from the latest report published by the Ministry for Ecological Transition [35] is shown in Table 2. All the fuel poverty indicators shown in this Table reveal that the percentages of households in Andalusia are higher than the same indicators at the national level. Consequently, formulating energy policies and acting on the energy efficiency of buildings in this region is essential.

In this context, this study aims to provide insight into the energy performance of the building stock in the most populated city of Andalusia, Seville. For this, the EPC database was selected, and the energy performance classification of residential buildings in Seville was exhaustively analysed. This study focusses on residential buildings (single-family dwellings and individual dwellings in multi-family blocks) and compares the results obtained based on different variables (type of building, construction period, and location). The bottom-up approach used in this study provides useful information to assess the energy performance of Seville's residential buildings at an urban scale and could support the policymaking process during the development of a trajectory for the progressive renovation of the residential building stock.

2 Methodology

2.1 Case of Study: Seville

The city of Seville, the capital of the Region of Andalusia, is located in southern Spain (coordinates: 37°23′00″N 5°59′00″W). The climate in Seville is Csa (hot-summer Mediterranean climate), according to the Köppen–Geiger classification. The average annual temperature is 18.8 °C, reaching average temperatures of 32.0 °C and 16.0 °C in summer and winter, respectively. Seville is a city with the largest population in Andalusia (681.998 inhabitants) [36] and has an urban surface area of 75 km^2. This city is divided into 11 districts (Fig. 1). Table 3 summarises the main characteristics of each district.

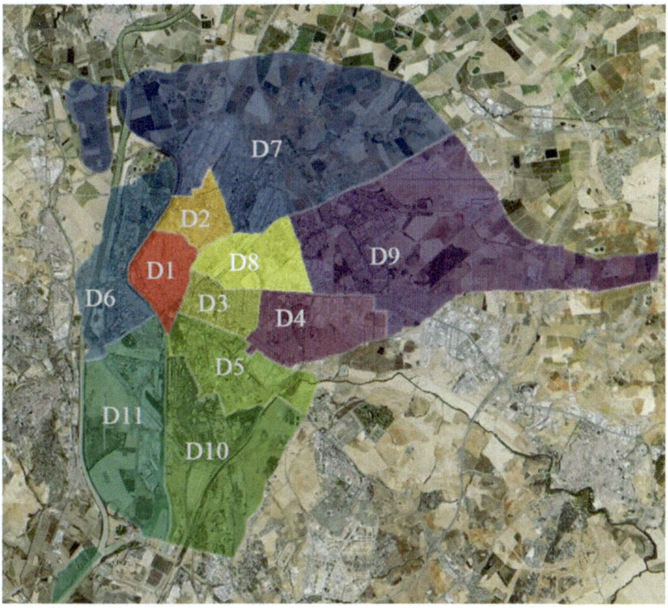

Fig. 1 Location of Seville's districts

Table 3 Characteristics of the districts of Seville

District	ID	Area (km^2)	Population	Population/area (km^{-2})
Casco antiguo	D1	4.23	57,068	13,491
Macarena	D2	3.17	73,682	23,244
Nervión	D3	3.20	51,084	15,964
Cerro-amate	D4	7.50	89,593	11,946
Sur	D5	7.49	70,274	9,382
Triana	D6	9.33	46,926	5,030
Norte	D7	38.57	71,235	1,847
San Pablo-Santa Justa	D8	5.63	59,068	10,492
Este	D9	31.38	106,490	3,394
Bellavista-Palmera	D10	16.04	42,284	2,636
Los Remedios	D11	15.56	25,525	1,640

2.2 Methods

The methodology used in this study is divided into three steps: (1) data extraction, (2) data pre-processing, and (3) data analysis. Figure 2 shows the research flowchart for this study.

Step 1 involved the extraction of data from different open Spanish governmental databases. The data used in this study have been extracted from (1) the Open Data Portal database [37], (2) the Cadastre database [38], and (3) the National Statistics Institute (NSI) database [39]. The information on EPCs was obtained from the database of the Open Data Portal [37]. Table 4 furnishes a summary of the extracted data.

The cadastral cartography in vector format (shapefile) was obtained from the Cadastre database [38]. Finally, information on the districts was obtained from the National Statistics Institute database [39].

Step 2 consisted of pre-processing the data obtained from the different databases. First, a data cleaning process was carried out. For this, missing data were identified and classified. EPCs with missing values were deleted. Subsequently, an analysis of noisy data (outliers) was performed to decide about their inclusion or exclusion in the final database. Duplicate values were identified and removed. Finally, a filtering of the data was performed by the city of Seville and by type of residential building (single-family dwelling and individual dwelling in a multi-family block).

Further, since each EPC corresponds to a single dwelling, in the case of single-family dwellings in a multi-family block, the values of global EP_{nr} consumption and global CO_2 emission cannot be used directly to evaluate the energy performance of an entire building. To address this limitation, a scaled-up index was used following the methodology applied by Conticelly et al. [40]. This index is based on a weighted

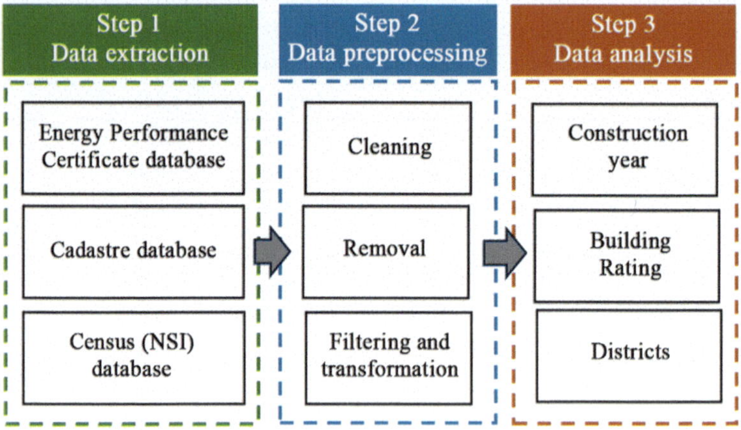

Fig. 2 Research flowchart

Table 4 Information compiled on EPCs

Information of the building	Address (municipality, postcode, province, and autonomous community)
	Climate zone
	Year of construction
	Regulations
	Cadastral reference
	Building type
Energetic performance of the building	Building identification
	General data and geometry
	EP_{nr} consumption (global, heating, cooling, and DHW)
	CO_2 emissions (global, heating, cooling, and DHW)
	Rating (energy efficiency and emission environmental impact)

average of the global EP_{nr} consumption and global CO_2 emission values of each dwelling in the same building. Equations 1 and 2 show the calculation process of the two values, respectively:

$$EP_{nr,\text{ average}} = \frac{\sum_{i=1}^{n} EP_{nr,i} \cdot S_i}{\sum_{i=1}^{n} S_i} \tag{1}$$

$$CO_{2,\text{ average}} = \frac{\sum_{i=1}^{n} CO_{2,i} \cdot S_i}{\sum_{i=1}^{n} S_i} \tag{2}$$

where $EP_{nr,\text{ average}}$ is the global EP_{nr} consumption average value of the entire building (kWh/m^2·year); $CO_{2,\text{ average}}$ is the global CO_2 emission average value of the entire building (kgCO_2/m^2·year); n is the total number of dwellings in the buildings whose EPCs are in the database; S_i is the floor area of the i-dwelling; $EP_{nr,i}$ is the global EP_{nr} consumption value of the i-dwelling, and $CO_{2,i}$ is the global CO_2 emission value of the i-dwelling.

Step 3 includes the statistical analysis of the database. Data were analysed based on the year of construction, the global EP_{nr} consumption, the global CO_2 emission, and the location by districts. The non-parametric Mann–Whitney U test was used to analyse the differences between dwellings in different districts. This statistical analysis was conducted using SPSS software (v. 29.0) [41]. In addition, QGIS (vs. 3.36.0) [42] software was used for data gathering and visualisation.

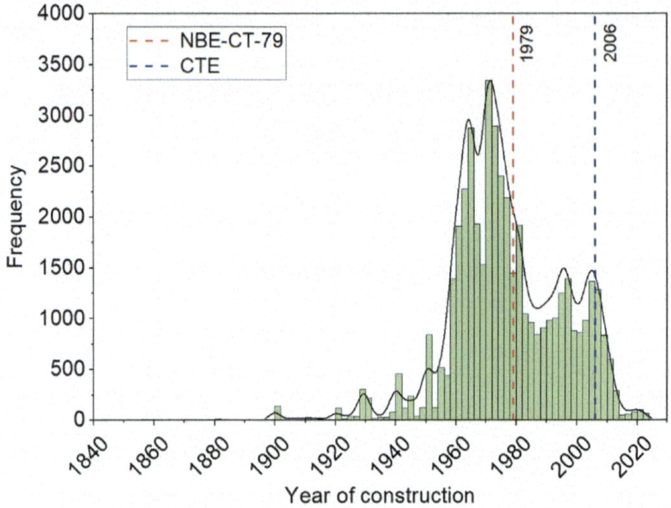

Fig. 3 Distribution of number of dwellings built per year obtained from EPC database

3 Results and Discussion

3.1 Year of Construction and Building Regulations

A total of 45,920 EPCs were obtained from the database: 40,677 individual dwellings in multi-family blocks and 5,243 single-family dwellings. The year of construction of these buildings ranges from 1850 to 2023. Figure 3 shows the distribution of the year of construction of the buildings obtained from the EPC database. According to the energy performance building regulations, 60% (27,437) are EPCs of dwellings built before 1979, 33% (15,124) relating to 1980–2006 and 7% (3,359) constructed after 2006. Figure 3 shows that the highest distribution of EPCs is located around 1970 and 1964. After the NBE-CT 79 regulation came into force, three concentrations were observed around the years 1980, 1996, and 2004.

3.2 Energy Performance

Table 5 summarises the energy performance results extracted from the EPCs by dwelling type. Global EP_{nr} consumption and global CO_2 emissions are found to be higher in single-family dwellings than in individual dwellings in multi-family blocks. Concerning EP_{nr} consumption, the system that consumes the largest proportion of energy in individual dwellings in multi-family blocks is the heating system (77.3 kWh/m^2·year), followed by the DHW production system (50.4 kWh/m^2·year) and the cooling system (30.1 kWh/m^2·year). In the case of single-family dwellings, the heating system is responsible for the highest energy consumption (98.1 kWh/m^2·year), while the DHW and cooling systems have similar consumptions (37.8 kWh/m^2·year and 32.9 kWh/m^2·year, respectively).

Similar trends are observed for CO_2 emissions. The heating system is responsible for most of the emissions in both individual dwellings in multi-family blocks (15.2 kgCO_2/m^2·year) and single-family dwellings (19.4 kgCO_2/m^2·year).

Figure 4 shows the average EP_{nr} consumption according to the system (i.e. heating, cooling, and DHW), type of building, and the energy performance regulation applicable when the dwelling was built. Generally, it is observed that EP_{nr} consumption decreases according to the different construction periods, with heating consumption decreasing the most. The decrease in this type of EP_{nr} consumption may be associated with the improved performance of opaque envelopes in terms of transmittance consequent to increased regulatory requirements. In contrast, there is minimal variation in EP_{nr} consumption due to cooling systems. A comparison by type of dwelling shows that EP_{nr} consumption in single-family dwellings has decreased according to the regulations, showing consumption values similar to those of individual dwellings in multi-family blocks built as of 2006 (period of entry into force of the CTE).

Concerning the energy performance rating, Figure 5 shows the distribution of EPCs according to rating and dwelling type. Most of the individual dwellings in a multi-family block are rated E (67%), followed by G (18%) and F (9%). In contrast, the top three ratings for single-family dwellings are E (66%), D (19%), and F (6%). Differences are observed between the lower and upper rating letters depending on the

Table 5 Summary of EP_{nr} consumption and CO_2 emissions in Seville according to EPCs

Type of building	Parameter		Mean	Median	S.D
Individual dwelling in a multi-family block	EP_{nr} consumption (kWh/m^2·year)	Global	157.8	151.4	53.5
		Heating	77.3	73.0	37.4
		Cooling	30.1	28.5	10.5
		DHW	50.4	43.5	30.8
	CO_2 emission (kgCO_2/m^2·year)	Global	30.0	29.0	10.1
		Heating	15.2	14.3	7.5
		Cooling	5.1	4.8	1.8
		DHW	9.7	8.9	5.4
Single-family dwellings	EP_{nr} consumption (kWh/m^2·year)	Global	168.9	164.7	60.6
		Heating	98.1	96.6	43.3
		Cooling	32.9	31.1	12.4
		DHW	37.8	33.5	28.3
	CO_2 emission (kgCO_2/m^2·year)	Global	32.3	31.4	11.9
		Heating	19.4	19.0	8.9
		Cooling	5.6	5.3	2.1
		DHW	7.3	6.7	5.2

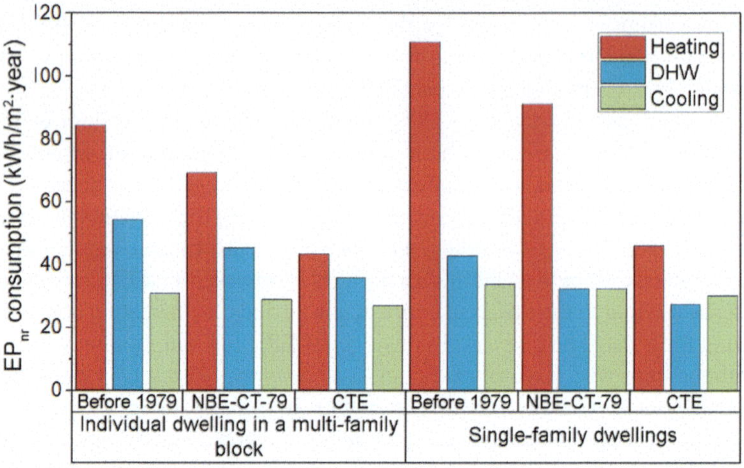

Fig.4 Average EP$_{nr}$ consumption by system, building type, and building regulations

type of building: while the letters G and F add up to 27% of the EPCs in individual dwellings in a multi-family block, the same categories add up to 9% of the total in single-family dwellings. On the other hand, the sum of the letters A, B, and C shows a percentage of less than 1% for individual dwellings in a multi-family block and around 5% for single-family dwellings.

Figure 6 shows the distribution of EPCs according to energy performance regulations and type of dwellings. For individual dwellings in a multi-family block, the values reveal that although most of the EPC are classified in category E, a slight improvement is observed during the period of application of the CTE (categories C and D increase to 7% and 28%, respectively).

In the case of single-family dwellings, a favourable trend is observed concerning improving energy categories as the energy efficiency regulation evolves. It stands out that most of the dwellings are classified in category E before 1979 (72%) and

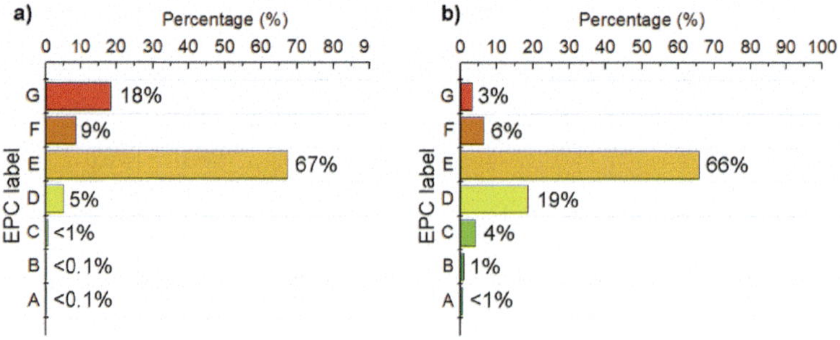

Fig. 5 Energy performance rating by type of building: **a** individual dwellings in a multi-family block and **b** single-family dwellings

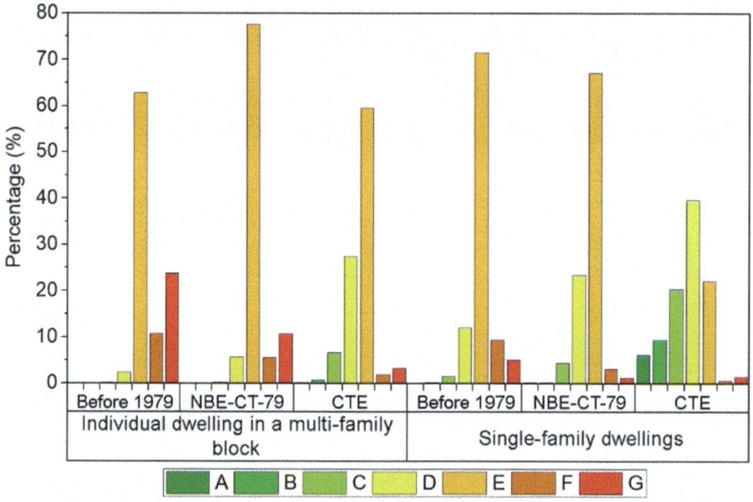

Fig. 6 Energy performance rating by type of building and building regulations

during the application of the NBE-CT-79 standard (67%). During the CTE application period, a higher percentage of dwellings fall into the most efficient categories (A, B, and C), indicating a significant improvement in energy efficiency under this standard.

3.3 Energy Performance by District

Regarding the number of dwelling EPCs in each district of Seville city, the values show that the distribution is not homogeneous. The districts with the highest number of dwelling EPCs are D4 (13.60%), D2 (13.44%), and D9 (12.52%). In contrast, the districts with a lower percentage of EPC are D11 (3.73%) and D3 (5.53%). Nevertheless, all districts are predominantly EPCs of individual dwellings in a multi-family block (Table 6). Only in districts D4, D9, and D10, the number of EPCs of single-family dwellings is higher than 20%, compared to EPCs of individual dwellings in a multi-family block.

Figure 7 shows the location of the dwellings with EPC in Seville and their year of construction. As expected, and notably, the percentage of dwellings built before 1979 is higher in districts D2 and D4, indicating that many dwellings in these districts were built before the implementation of the NBE-CT-79 standard. During the period of application of this standard, the number of EPCs decreased significantly in most districts, compared to the previous period. However, districts D7 and D9 show an increase in the number of certificates in this period. In contrast, districts D9 and D10 show a higher number of EPCs during the CTE application period, compared to other districts. This fact could indicate an increase in the construction of newer dwellings in these areas.

Table 6 Summary of EPCs distribution by district and building type

District	Individual dwelling in a multi-family block	Single-family dwellings
D1	4,335 (90%)	473 (10%)
D2	6,043 (98%)	129 (2%)
D3	2,973 (89%)	381 (11%)
D4	4,832 (77%)	1,412 (23%)
D5	4,290 (96%)	163 (4%)
D6	3,100 (95%)	173 (5%)
D7	3,001 (93%)	218 (7%)
D8	4,011 (91%)	382 (9%)
D9	4,469 (78%)	1,279 (22%)
D10	1,960 (77%)	581 (23%)
D11	1,663 (97%)	52 (3%)

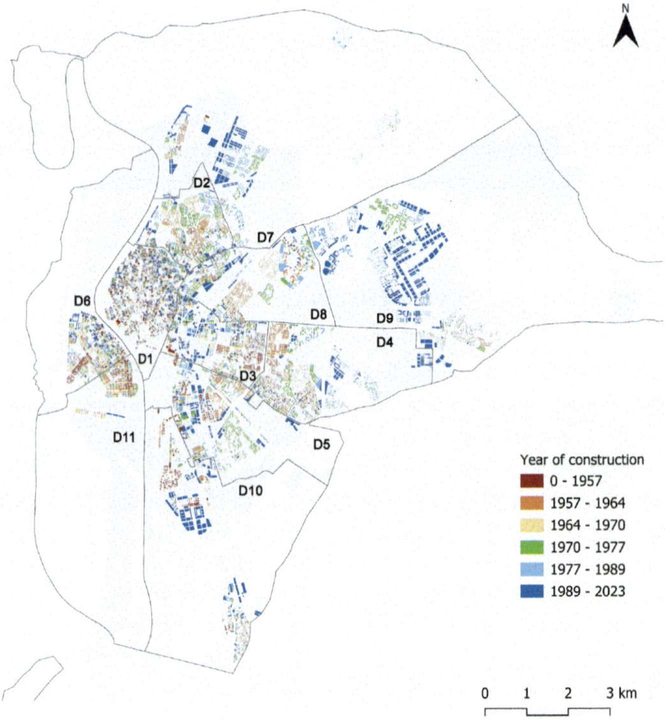

Fig. 7 Year of construction of buildings in Seville. High resolution available in [43]

Regarding average EP_{rn} consumption in individual dwellings in multi-family blocks, Figure 8a summarises the values obtained. Districts D4 (171 kWh/m^2·year) and D2 (169 kWh/m^2·year) show the highest mean values of EP_{rn} consumption, while D9 has the lowest mean consumption (139 kWh/m^2·year). Quartile analysis results in Fig. 8a indicate that the first quartile varies significantly between districts, with districts D2 and D4 showing the highest values, revealing that even the lowest consuming dwellings in these districts have relatively high consumption. The third quartile shows that districts D4 and D2 have the highest values, which is consistent with their high average consumption. Regarding consumption variability, district D1 has the highest interquartile range (IQR), indicating greater variability in EP_{rn} energy consumption, while D9 shows the lowest IQR, indicating that the average EP_{rn} energy consumption is more uniform across dwellings. A slightly different distribution is observed for single-family dwellings (Fig. 8a). The district with the highest average consumption is D4 (185 kWh/m^2·year) while D1 (155 kWh/m^2·year) has the lowest consumption. Interquartile range analysis shows that D9 has the lowest interquartile range (64.3 kWh/m^2·year) and D3 the highest (88.7 kWh/m^2·year). In contrast to individual dwellings in a multi-family block, the highest interquartile range is observed in districts D3 and D10, while the highest average consumption is found in D4 and D1. Regarding CO_2 emissions, similar patterns are observed between districts (Fig. 8b).

In addition, a mapping of the average EP_{nr} consumptions and CO_2 emissions per dwelling and district is shown in Figs. 9 and 10, respectively. The high-resolution maps can be found in Guerrero-Rivera et al. [43]

Statistically significant differences in the global EP_{rn} consumption and the global CO_2 emissions of dwellings between districts were also evaluated. Table 7 shows the results of the Mann–Whitney *U* tests between districts from individual dwellings in multi-family blocks. As for consumption, the results reveal significant differences between most districts, except between districts D1, D5, D6, and D8 and between the

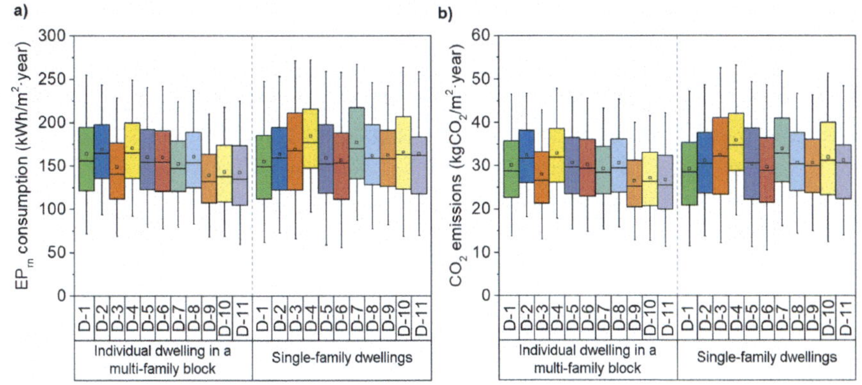

Fig. 8 **a** EP_{nr} consumption by district and type of dwelling. **b** CO_2 emissions by district and type of dwelling

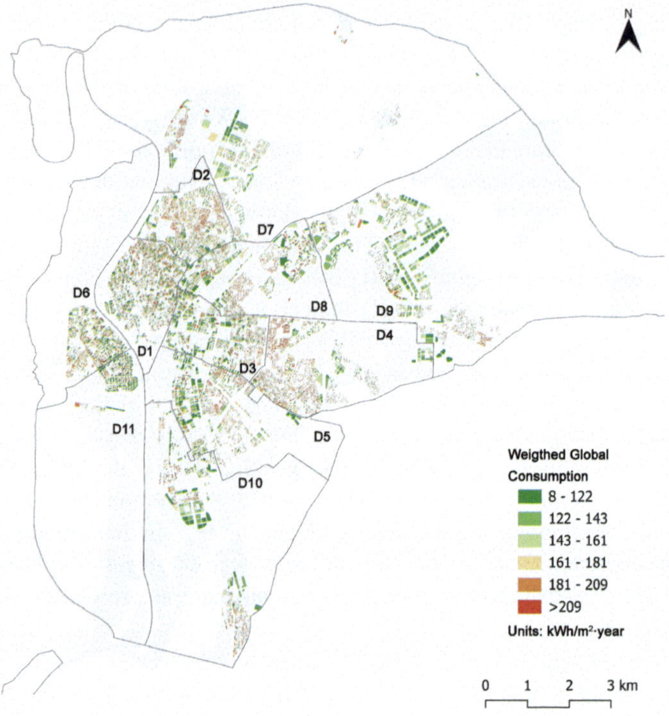

Fig. 9 Distribution of the weighted global EP_{nr} consumptions of residential buildings in Seville. High resolution available in [43]

districts with the highest average consumption (D2 and D4). Regarding global CO_2 emissions, the values also show significant differences between all districts except between D5 and D6, and D2 and D4. In contrast, the results obtained for single-family dwellings (Table 8) show a greater diversity of results, with no significant differences in global consumption and global CO_2 emissions among more districts.

Finally, Figure 11a and b shows dwelling classification according to the EP_{nr} consumption label and global CO_2 emissions label, respectively. Regarding EP_{nr}, the worst-energy-performing districts are districts D4 (36%) and D2 (34%). Category E is where the most dwellings are classified in all districts, both in terms of EP_{nr} label and CO_2 emissions label.

A mapping of the EPC rating per category can be found in Guerrero-Rivera et al. [43].

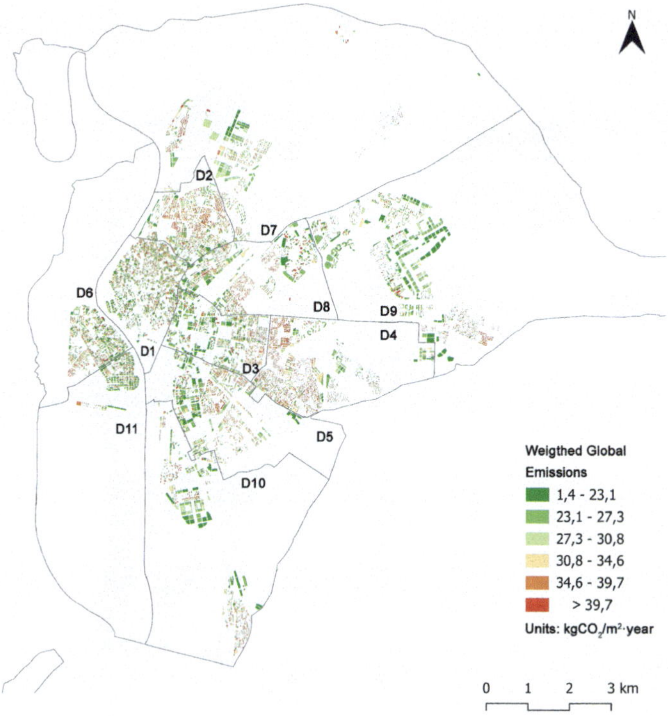

Fig. 10 Distribution of the weighted global emissions of residential buildings in Seville. High resolution: Guerrero-Rivera et al. [43]

4 Conclusions

Residential buildings are responsible for a large share of energy consumption in cities. This study analysed the energy performance of the residential building stock in Seville, the capital and most populated city of Andalusia. A total of 45,920 EPCs were processed and analysed. The EPC database shows a varied stock of residential buildings in Seville, with construction ranging over a wide period (1850–2023). Most of the residential building EPCs registered in the public database (60%) correspond to buildings built before 1979, i.e. before the entry into force of the first energy performance building regulation in Spain.

The results obtained showed that most of the dwellings show a poor energy performance rating (approximately 66% are classified under label E). The average global EP_{nr} consumption of individual dwellings in a multi-family block was 157.8 kWh/m^2·year, with this value rising to 168.9 kWh/m^2·year for single-family dwellings.

In addition, the results revealed that the districts with the highest percentage of worst-energy-performing dwellings (i.e. dwellings with EPC "G" and "F") are districts D4 (36%) and D2 (34%). Elaborating policies focused on boosting building renovations in these districts could significantly transform Seville's residential

Table 7 Results obtained from the Mann–Whitney U test to determine the significant differences between the global EP_{nr} consumption and global CO_2 emission in individual dwellings in multi-family blocks obtained between districts. Districts with statistically significant differences are put in bold (p-value < 0.05)

	D1	D2	D3	D4	D5	D6	D7	D8	D9	D10
Global EP_{nr} consumption										
D2	**1.2E + 07**									
D3	**7.4E + 06**	**1.1E + 07**								
D4	**9.3E + 06**	1.4E + 07	**5.3E + 06**							
D5	9.4E + 06	**1.4E + 07**	**5.5E + 06**	**1.2E + 07**						
D6	6.8E + 06	**1.0E + 07**	**4.0E + 06**	**8.5E + 06**	6.7E + 06					
D7	**7.1E + 06**	**1.1E + 07**	**4.2E + 06**	**8.8E + 06**	**7.0E + 06**	**5.0E + 06**				
D8	8.8E + 06	**1.4E + 07**	**5.1E + 06**	**1.1E + 07**	8.6E + 06	6.2E + 06	**5.5E + 06**			
D9	**1.2E + 07**	**1.8E + 07**	**7.3E + 06**	**1.5E + 07**	**1.2E + 07**	**8.5E + 06**	**7.9E + 06**	**1.1E + 07**		
D10	**5.1E + 06**	**7.7E + 06**	**3.0E + 06**	**6.2E + 06**	**5.0E + 06**	**3.6E + 06**	**3.3E + 06**	**4.7E + 06**	**4.1E + 06**	
D11	**4.4E + 06**	**6.6E + 06**	**2.7E + 06**	**5.3E + 06**	**4.3E + 06**	**3.1E + 06**	**2.8E + 06**	**4.1E + 06**	3.6E + 06	1.7E + 06
Global CO_2 emission										
D2	**1.1E + 07**									
D3	**7.2E + 06**	**1.2E + 07**								
D4	**8.6E + 06**	1.4E + 07	**5.0E + 06**							
D5	**8.9E + 06**	**1.5E + 07**	**5.3E + 06**	**1.2E + 07**						
D6	6.6E + 06	**1.1E + 07**	**4.0E + 06**	**8.7E + 06**	6.8E + 06					
D7	6.6E + 06	**1.1E + 07**	**4.0E + 06**	**8.8E + 06**	**6.9E + 06**	**4.9E + 06**				
D8	**8.3E + 06**	**1.4E + 07**	**5.0E + 06**	**1.1E + 07**	8.6E + 06	6.1E + 06	**5.6E + 06**			
D9	**1.2E + 07**	**1.9E + 07**	**7.2E + 06**	**1.5E + 07**	**1.2E + 07**	**8.5E + 06**	**8.0E + 06**	**1.1E + 07**		

(continued)

Table 7 (continued)

	D1	D2	D3	D4	D5	D6	D7	D8	D9	D10
D10	**4.9E + 06**	**7.8E + 06**	**3.0E + 06**	**6.3E + 06**	**5.0E + 06**	**3.6E + 06**	**3.3E + 06**	**4.7E + 06**	**4.2E + 06**	
D11	**4.3E + 06**	**6.8E + 06**	**2.7E + 06**	**5.5E + 06**	**4.4E + 06**	**3.1E + 06**	**2.9E + 06**	**4.1E + 06**	3.7E + 06	**1.7E + 06**

building stock. By decreasing the average EP_{nr} consumption and renovating the most energy-vulnerable areas, Seville can move towards a zero-emission building stock. Measures such as technical assistance and information campaigns, especially targeting districts with poor energy-performing buildings, would contribute to this transformation.

Nevertheless, it is important to note that this study relies on data from the open EPC database. As this database continues to expand, including new EPCs for both newly constructed buildings and those previously lacking EPCs, future studies will be necessary. These studies can evaluate the evolution of the city's building stock renovation strategies and their effectiveness in reducing the city's average EP_{nr} consumption.

Table 8 Results obtained from the Mann–Whitney U test to determine the significant differences between the global EP_{nr} consumption and global CO_2 emission in single-family dwellings obtained in each district. Districts with statistically significant differences are put in bold. (p-value < 0,05)

	D1	D2	D3	D4	D5	D6	D7	D8	D9	D10
Global EP_{nr} consumption										
D2	2.7E + 04									
D3	**7.7E + 04**	2.3E + 04								
D4	**2.3E + 05**	**7.1E + 04**	**2.3E + 05**							
D5	3.6E + 04	1.1E + 04	3.4E + 04	**1.4E + 05**						
D6	4.0E + 04	1.2E + 04	**3.7E + 04**	**1.6E + 05**	1.5E + 04					
D7	**4.0E + 04**	**1.2E + 04**	3.9E + 04	1.7E + 05	**1.5E + 04**	**1.5E + 04**				
D8	**8.1E + 04**	2.4E + 04	7.7E + 04	**3.3E + 05**	3.0E + 04	3.0E + 04	**4.7E + 04**			
D9	**2.7E + 05**	8.3E + 04	**2.6E + 05**	**1.1E + 06**	1.0E + 05	1.0E + 05	**1.6E + 05**	2.5E + 05		
D10	**1.2E + 05**	3.6E + 04	1.1E + 05	**4.8E + 05**	4.4E + 04	**4.5E + 04**	**6.9E + 04**	1.1E + 05	3.5E + 05	
D11	1.1E + 04	3.4E + 03	1.1E + 04	**4.5E + 04**	4.2E + 03	4.2E + 03	6.5E + 03	1.0E + 04	3.4E + 04	1.6E + 04
Global CO_2 emissions										
D2	**2.7E + 04**									
D3	**7.6E + 04**	2.3E + 04								
D4	**2.2E + 05**	**6.8E + 04**	**2.3E + 05**							
D5	3.6E + 04	1.1E + 04	3.4E + 04	**1.5E + 05**						
D6	4.0E + 04	1.2E + 04	**3.7E + 04**	**1.6E + 05**	1.5E + 04					
D7	**3.9E + 04**	**1.2E + 04**	3.9E + 04	**1.7E + 05**	**1.5E + 04**	**1.5E + 04**				
D8	**8.1E + 04**	2.5E + 04	7.8E + 04	**3.4E + 05**	3.0E + 04	3.1E + 04	**4.8E + 04**			
D9	**2.7E + 05**	8.5E + 04	**2.7E + 05**	**1.2E + 06**	1.0E + 05	1.0E + 05	**1.6E + 05**	2.5E + 05		

(continued)

Table 8 (continued)

	D1	D2	D3	D4	D5	D6	D7	D8	D9	D10
D10	**1.2E + 05**	3.6E + 04	1.1E + 05	**4.9E + 05**	4.4E + 04	**4.5E + 04**	6.9E + 04	1.1E + 05	**3.4E + 05**	
D11	1.1E + 04	3.4E + 03	1.1E + 04	**4.7E + 04**	4.1E + 03	4.2E + 03	6.5E + 03	1.0E + 04	3.3E + 04	1.6E + 04

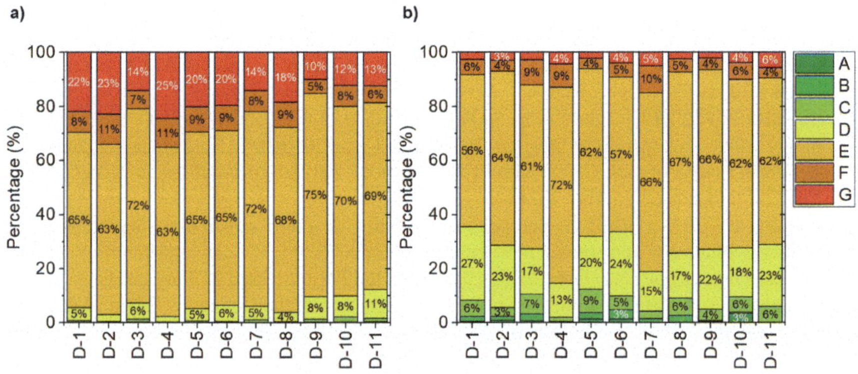

Fig.11 Distribution of energy performance rating in **a** individual dwellings in multi-family blocks by district and **b** single-family dwellings by district

Acknowledgements María Luisa de la Hoz-Torres wishes to acknowledge the support of the MICIU/AEI/https://doi.org/10.13039/501100011033 and by European Union NextGeneration EU/ PRTR under a Juan de la Cierva post-doctoral contract (JDC2022-049561-I).

References

1. In focus: Energy efficiency in buildings—European Commission (2020) https://commission.europa.eu/news/focus-energy-efficiency-buildings-2020-02-17_en (accessed May 8, 2024)
2. European Commission (2024) EU energy statistical pocketbook and country datasheets, (2024). https://energy.ec.europa.eu/data-and-analysis/eu-energy-statistical-pocketbook-and-country-datasheets_en (accessed May 16, 2024)
3. Valenzuela-Flores A, Marín-Restrepo L, Pereira-Ruchansky L, Pérez-Fargallo A (2023) Impact of energy literacy on vulnerable families: case study—the chilean good energy (Con Buena Energía) program. Los Ríos Region Energy Policy 180:113650. https://doi.org/10.1016/J.ENPOL.2023.113650
4. International Energy Agency, Heating (2023) https://www.iea.org/energy-system/buildings/heating (accessed 8 May 2024)
5. Aguilar AJ, de la Hoz-Torres ML, Costa N, Arezes P, Martínez-Aires MD, Ruiz DP (2023) Indoor acoustic quality of educational buildings in South West Europe: influence of current ventilation strategies. J Build Eng 80:108012. https://doi.org/10.1016/J.JOBE.2023.108012

6. de la Hoz-Torres ML, Aguilar AJ, Ruiz DP, Martínez-Aires MD (2024) An investigation of indoor thermal environments and thermal comfort in naturally ventilated educational buildings. J Build Eng 84:108677. https://doi.org/10.1016/J.JOBE.2024.108677
7. Adaptive F, Delgado-Gutierrez E, Canivell J, Bienvenido-Huertas D, Hidalgo-Sánchez FM (2024) Adaptive comfort potential in different climate zones of Ecuador considering global warming. Energies. https://doi.org/10.3390/EN17092017
8. Bienvenido-Huertas D, de la Hoz-Torres ML, Aguilar AJ, Tejedor B, Sánchez-García D (2023) Holistic overview of natural ventilation and mixed mode in built environment of warm climate zones and hot seasons. Build Environ 245:110942. https://doi.org/10.1016/J.BUILDENV.2023.110942
9. Directive 2010/31/EU of the European Parliament and of the Council of 19 May 2010 on the energy performance of buildings, (2010). https://eur-lex.europa.eu/eli/dir/2010/31/oj (accessed 25 May 2024)
10. Directive 2012/27/EU of the European Parliament and of the Council of 25 October 2012 on energy efficiency, amending Directives 2009/125/EC and 2010/30/EU and repealing Directives 2004/8/EC and 2006/32/EC Text with EEA relevance, (2012). https://eur-lex.europa.eu/eli/dir/2012/27/oj (accessed 25 May 2024)
11. Directive EU 2023/1791 of the European Parliament and of the Council of 13 September 2023 on energy efficiency and amending Regulation (EU) 2023/955, (2023). https://eur-lex.europa.eu/legal-content/EN/TXT/?uri=CELEX%3A32023L1791 (accessed 25 May 2024)
12. Directive EU 2024/1275 of the European Parliament and of the Council of 24 April 2024 on the energy performance of buildings (2024) https://eur-lex.europa.eu/legal-content/EN/TXT/?uri=CELEX%3A32024L1275 (accessed 25 May 2024)
13. European Commission, Energy Performance of Buildings Directive (2024) https://energy.ec.europa.eu/topics/energy-efficiency/energy-efficient-buildings/energy-performance-buildings-directive_en (accessed 9 May 2024)
14. European Commission, The European Green Deal (2024) https://commission.europa.eu/strategy-and-policy/priorities-2019-2024/story-von-der-leyen-commission/european-green-deal_en (accessed 23 May 2024)
15. United Nation, THE 17 GOALS | Sustainable Development (2024) https://sdgs.un.org/goals (accessed 23 May 2024)
16. United Nation, The Sustainable Development Goals Report. Special edition (2023) https://unstats.un.org/sdgs/report/2023/The-Sustainable-Development-Goals-Report-2023.pdf (accessed 8 May 2024)
17. Green Building Council España, Informe País GBCe 2022: sobre el estado de las urgencias de la edificación sostenible en España
18. Orden de 20 de mayo de 1960 por la que se aprueba la adaptación de las ordenanzas técnicas y normas constructivas, aprobadas por Ordenes de 12 de julio de 1955 y 22 de febrero de 1968, al texto refundido y revisado de la Legislación de Viviendas de Protección Oficial y su Reglamento (1960) https://www.boe.es/boe/dias/1969/05/23/pdfs/A07918-07923.pdf (accessed 6 June 2024)
19. Real Decreto 2177/1996, de 4 de octubre, por el que se aprueba la Norma Básica de la Edificación "NBE-CPI/96: condiciones de protección contra incendios de los edificios, (1996). https://www.boe.es/boe/dias/1996/10/29/pdfs/A32378-32422.pdf (accessed 6 June 2024)
20. Documento Básico "Ahorro de Energía" del Código Técnico de la Edificación (Versión 2006), (n.d.). https://www.codigotecnico.org/pdf/Documentos/HE/DBAnteriores/DBHE_200602.pdf (accessed 6 June 2024)
21. Decreto 47/2007, de 19 de enero, por el que se aprueba el Procedimiento básico para la certificación de eficiencia energética de edificios de nueva construcción (2007) https://www.boe.es/buscar/doc.php?id=BOE-A-2007-2007 (accessed 6 June 2024)
22. Real Decreto 235/2013, de 5 de abril, por el que se aprueba el procedimiento básico para la certificación de la eficiencia energética de los edificios. (2013) https://www.boe.es/buscar/act.php?id=BOE-A-2013-3904 (accessed 6 June 2024)

23. Documento Básico "Ahorro de energía" del Código Técnico de la Edificación. (Versión 2013) (n.d.) https://www.codigotecnico.org/pdf/Documentos/HE/DBAnteriores/DBHE_20131108.pdf (accessed 6 June 2024)
24. M and UA Ministry of Transport, Guías de aplicación del DB-HE 2019, (2022) https://www.codigotecnico.org/pdf/GuiasyOtros/Conceptos_basicos_RD_732_2019.pdf (accessed 6 June 2024)
25. European Parliament (2024) Revision of the energy performance of buildings directive-Q4 2021
26. Santini L (2021) Next steps for MEPS: designing minimum energy performance standards for Europe, Regulatory Assistance Project https://policycommons.net/artifacts/3450819/next-steps-for-meps/4251076/ (accessed 26 May 2024)
27. Jiménez Torres M, May Tzuc O, Vázquez-Torres CE, Bassam A (2024) Mapping of building energy consumption and emissions under representative concentration pathway scenarios by a geographic information system descriptive framework: case study of Mexico. Sustain Energy Plan Smart Grids. https://doi.org/10.1016/B978-0-443-14154-6.00010-7
28. May Tzuc O, Jiménez Torres M, Rodriguez CM, Demesa López FN, Noh Pat F (2023) Cluster analysis as a tool for the territorial categorization of energy consumption in buildings based on weather patterns. Stud Comput Intell 1105:73–91. https://doi.org/10.1007/978-3-031-37454-8_4/TABLES/5
29. Fabbri K, Gaspari J (2021) Mapping the energy poverty: a case study based on the energy performance certificates in the city of Bologna. Energy Build 234:110718. https://doi.org/10.1016/J.ENBUILD.2021.110718
30. Li W, Chien F, Hsu CC, Zhang YQ, Nawaz MA, Iqbal S, Mohsin M (2021) Nexus between energy poverty and energy efficiency: estimating the long-run dynamics. Resour Policy 72:102063. https://doi.org/10.1016/J.RESOURPOL.2021.102063
31. Bienvenido-Huertas D, Sánchez-García D, Marín-García D, Rubio-Bellido C (2023) Is the analysis scale crucial to assess energy poverty? analysis of yearly and monthly assessments using the 2 M indicator in the south of Spain. Energy Build 285:112889. https://doi.org/10.1016/J.ENBUILD.2023.112889
32. Bienvenido-Huertas D, Sánchez-García D, Rubio-Bellido C, Marín-García D (2024) Holistic analysis to reduce energy poverty in social dwellings in southern Spain considering envelope, systems, operational pattern, and income levels. Energy 288:129796. https://doi.org/10.1016/J.ENERGY.2023.129796
33. Eurostat (2024) EU statistics on income and living conditions (EU-SILC), https://doi.org/10.2908/ILC_MDES01
34. Clavijo-Núñez S, Herrera-Limones R, Rey-Pérez J, Torres-García M (2022) Energy poverty in Andalusia. Anal Through Decentralised Ind Energy Policy 167:113083. https://doi.org/10.1016/J.ENPOL.2022.113083
35. Ministry for Ecological Transition and the Demographic Challenge (2022) Actualización de indicadores de la estrategia nacional contra la pobreza energética. Diciembre de 2022
36. Multi-territorial Information System of Andalusia (SIMA) (2022) Population of the municipality Sevilla https://www.juntadeandalucia.es/institutodeestadisticaycartografia/sima/nucleos.htm?CodMuni=41091 (accessed May 23, 2024)
37. Junta de Andalucía (2024) Databases of Energy Certificates of Andalusia, https://datos.gob.es/en/catalogo/a01002820-registro-de-certificados-energeticos-de-andalucia (accessed May 8, 2024)
38. Ministry of Finance of Spain, Cadastral Electronic Site, (n.d.) https://www.sedecatastro.gob.es/ (accessed May 8, 2024).
39. National Statistics Institute of Spain (INE) (2022) Population and Housing Censuses. Latest data, https://www.ine.es/dyngs/INEbase/es/operacion.htm?c=Estadistica_C&cid=1254736176992&menu=ultiDatos&idp=1254735572981 (accessed May 8, 2024)
40. Conticelli E, Falcioni S, Marzani G, Morini GL, Tondelli S (2024) Assessing energy efficiency at urban scale through the use of energy performance certificates: an application in the Emilia-Romagna region. Italy Cities 145:104728. https://doi.org/10.1016/J.CITIES.2023.104728

41. IBM (2024) SPSS Statistics, https://www.ibm.com/es-es/products/spss-statistics (accessed May 25, 2024)
42. QGIS (2024) Free and open source geographic information system, https://www.qgis.org/en/site/ (accessed May 25, 2024)
43. Guerrero-Rivera MF, Aguilar AJ, Aguilar-Camacho J, de la Hoz-Torres ML (2021) Mapping energy performance in Seville's existing residential buildings stock, Mapping Energy Performance in Sevilles Existing Residential Buildings Stock. https://doi.org/10.12795/11441/159928

Thermal Comfort in the Tropical Urban Context: A Study of Residential Housing in Ho Chi Minh City

Van Khang Huynh and Jesús Alberto Pulido Arcas

Abstract This chapter delves into the critical issue of adaptive thermal comfort in tropical countries, with a particular focus on contexts such as Vietnam, where a thriving economy is bringing new lifestyles associated with more energy-dependent air conditioning systems. On-site monitoring and a survey distributed to residents were used to collect information on indoor environment and adaptive comfort. Overall, our main results show that indoor temperatures of about 28 °C and relative humidity of 75% can be comfortable for the majority of residents and that air conditioners are rarely used. In conclusion, our findings emphasize the importance of developing thermal comfort models adapted to local conditions and cultural backgrounds.

1 Introduction

Recently, there has been a worldwide recognition of the problems associated with the environmental impact of the building sector: Reducing the carbon footprint, achieving energy neutrality, being resilient to energy insecurity, providing comfortable conditions during a climate crisis… As pointed out by Guy and Farmer [1], the broad concept of sustainability can be, and indeed is, approached from a plurality of perspectives. In the last decades, a more culturally sensible approach, "eco-cultural", has gained momentum, which places great emphasis on human comfort inside buildings from a local and cultural perspective. In addition, seminal authors in this field have also linked comfort to a *state of mind* that expresses satisfaction with the thermal environment ASHRAE (2009), as cited in [2], and consequently, considerable evidence has accumulated to show that this state of mind is dependent on cultural norms and practices.

V. K. Huynh (✉)
Institute of Smart City and Management, University of Economics Ho Chi Minh City, Ho Chi Minh City, Vietnam
e-mail: hv.khang@hutech.edu.vn

J. A. Pulido Arcas
Center for Global Education, The University of Tokyo, Tokyo, Japan

M. Torres González and C. Rubio Bellido (eds.), *Towards Low and Positive Energy Buildings*, Green Energy and Technology, https://doi.org/10.1007/978-3-031-70851-0_6

This chapter presents a study on thermal comfort from a culturally sensible perspective in Ho Chi Minh City in Vietnam, a country that has been experiencing profound cultural transformations, driven by a thriving economy and an exodus from rural to urban areas. The GDP per capita has almost multiplied by 10 in the last 20 years [3], and the percentage of urban population which reached 40,9% in 2019 has been fueled by high rates of internal migration, which accounted for as much as 8.6% of the total Vietnamese workforce in the same year [4]. These drastic changes on the socioeconomic structure of the country have also entailed changes in the way in which Vietnamese new urbanites dwell.

Specifically in Ho Chi Minh City, this context is materialized in a residential landscape in transition between the vernacular and the modern. According to data available in 2019, two housing typologies are predominant in this city: The townhouse (also known as tube house, shop house, street house, laneway house, horse-gut house, lot house, and row house) accounts for 94.4% of the housing stock, and apartment buildings account for 5.2% [5]. Nonetheless, apartment buildings are becoming the predominant typology for new residents, and their number is on the rise. In contrast to the traditional shop houses, which are integrated into the dense and varied urban fabric and therefore have little access to sunlight and ventilation, the new condominium complexes have emerged in the form of high-rise towers with abundant access to natural ventilation and sunlight; in fact, their average number of stories has increased from only 3–4 in 1975, to more than 20 after the year 2000, and, in addition, their internal layout and flexibility are remarkably different from the traditional Vietnamese housing typologies [6].

Changes in housing typologies and cultures are also accompanied by changes in thermal comfort conditions, the latter also influenced by new construction standards and climate control equipment. This is particularly important in tropical climates, such as Ho Chi Minh City, with an average humidity of around 80% and an average temperature of around 28 °C throughout the year. In the traditional static comfort model, these conditions would result in high levels of discomfort, making it necessary the use of air conditioners. However, as a growing body of literature shows, understanding comfort as a "state of mind" implies that many other factors come into play to shape this notion of comfort. In recent years many studies have explored these different "comforts" that exist in a moment in time, in a concrete place, and in a specific culture, with a focus on the countries of the Global South, where a more culturally and environmentally sensitive understanding of comfort appears to be possible.

This chapter seeks to gain a better understanding of thermal comfort in a context of transition by exploring the relationship between indoor environmental conditions, adaptive measures, and perceptions of comfort in two selected dwellings in Ho Chi Minh City: A traditional town house and a new apartment. The data presented here were originally collected as part of the main author's master thesis' [7], although the data analysis and the main conclusions adopt a new perspective, not present in the original document, focusing on two main aspects: Indoor conditions and adaptive measures for thermal comfort.

Despite our results are limited to only two housing units, they can add to the rapidly expanding field of adaptive thermal comfort in tropical latitudes and thus help build a larger body of knowledge to assist these countries in achieving their own conceptualization of thermal comfort, specifically adapted to their cultural norms.

2 Methodology

Various methods were utilized to capture the different facets of adaptive thermal comfort. First, environmental variables were monitored in two representative house typologies, and the data were analyzed with respect to the two main comfort standards ASHRAE-55 and EN-16798. Second, a survey was distributed to 128 residents to clarify the adaptive measures to achieve comfort in their homes.

2.1 House Typologies. Tube Houses

The economic reforms of 1986 have brought great prosperity to Vietnam, but the urban fabric, especially in large urban agglomerations, has suffered the consequences of a large influx of migrant populations and a lack of urban regulation. The tube house is an example of this. This typology has been shaped by an additive process in which newly affluent families could afford to buy land in the city to build their own houses; sometimes the land was already occupied by traditional rural houses, which were demolished to make way for the new structure. With the increase in population density and economic activity, these houses grew in height.

The current typology is therefore the result of a process of transformation. Tube houses have an average of 4–5 floors and fit into a very narrow lot, with a front that is only about 5 m wide and a length of about 17–20 m. Each house has a main facade, usually facing a commercial street, and a back facade facing a back alley. Each house belongs to one family and combines different uses, stacked on 4–5 levels (Fig. 1), with a total gross area of about 400–500 square meters. The house is entered through the main living room, located on the ground floor, with no transitions; the kitchen and dining room are located at the back, connected to a small rear courtyard. The first and second floors contain the bedrooms for the family members. The third floor is half built, with an open terrace facing the main street and a laundry room, a storage room and an altar at the back. All levels are connected by a central staircase that is lit and ventilated by a skylight protruding from the roof.

In terms of construction materials, these houses are usually built with reinforced concrete and are walled with concrete blocks or bricks; there is no thermal insulation due to the aforementioned lack of building regulations [8]. The main façade is the element through which the house presents itself to the public, the visible face of the family, and is therefore a showcase of divergent styles: Traditional, postmodern,

Fig. 1 Photo of a row house (left) and a condominium (right)

contemporary, neoclassical, pseudo-traditional... The windows in the main façade are usually large in size to take advantage of the sunlight; setbacks and terraces are sometimes present as part of the design.

2.2 Selection of Housing Typologies. Apartment Building

Large condominium complexes have shaped the recent urban development in Ho Chi Minh City, usually as part of large-scale urban developments, where the residents can enjoy high-quality amenities [9]. The vertical layout of these complexes consists of a ground floor dedicated to parking, the first and second floors housing common services for the residents (e.g., kindergarten, supermarkets, convenience stores, common lounges, or administrative offices), and residential units located on the third floor and above. A common layout is usually square or rectangular in shape, with multiple protrudes to maximize the façade length and profit from sunlight and ventilation. It consists of three to four units on each floor, surrounding a central core with the staircase and the elevators. The typical catalogue includes 1-bedroom, 2-bedrooms, and 3-bedrooms apartments, for single homeowners or couples, small and large families (Fig. 2). The towers are built in reinforced concrete and can be as high as 20–30 stories.

2.3 Monitoring of Environmental Variables

The first phase of this study collected data for two main environmental variables, namely dry bulb temperature (DBT) and relative humidity (RH) in two different

Fig. 2 Plans of a row house (left) and a condominium (right)

residential typologies in Ho Chi Minh City: A detached traditional house, and a condominium unit. Multiple units of data loggers were installed in different rooms of both houses that are representative of the living conditions (living spaces, bedrooms and kitchens): Six sensors in the apartment, with a gross are of around 90 square meters, (Fig. 3) and seven in the detached house, with a gross area of around 250 square meters (Fig. 4). The measurements were carried out in two days, from August 17 to August 18 in the condominium, and from August 22 to August 23 in the apartment. In both cases, a 24-h cycle was measured in steps of 10 min.

2.4 *Survey Design*

The survey gathered information about 128 residents, who were surveyed in three main areas: Occupancy profiles, energy use patterns, and adaptive measures for thermal comfort.

Regarding occupancy profiles, the survey gathered data about the number of people in the house, both daily and weekly, and information about the house typology. Energy use patterns were clarified through information about the use of air conditioners; logically, heating was not included as it is not necessary in a tropical climate. The adaptive measures for thermal comfort were clarified with questions about different strategies for shading and natural ventilation. A full transcript of the questionnaire is included in [7].

Fig. 3 Sensor location in the apartment unit, from [7] (Reprinted with permission)

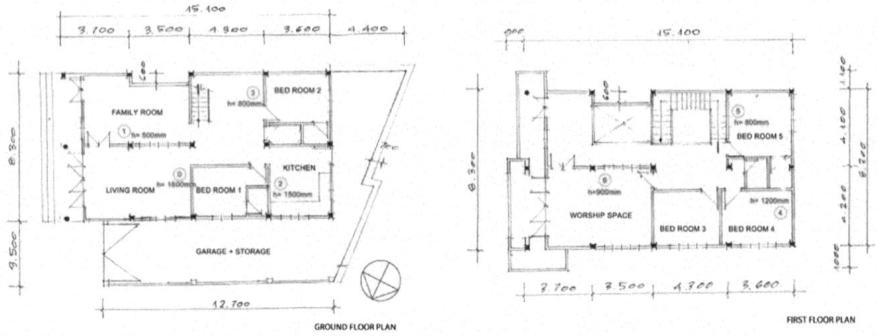

Fig. 4 Sensor location in the detached house, from [7] (Reprinted with permission)

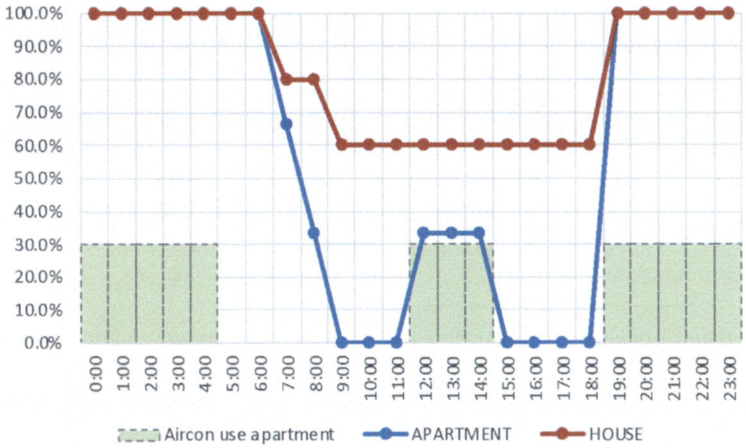

Fig. 5 Occupancy profiles and air conditioner use in a 24-h cycle

3 Results and Discussion

3.1 Occupancy Profiles and Energy Use in the Two Selected Typologies

Three members of the same family (father, son, and daughter) inhabit the apartment and five members (father, mother, son, son's wife, and daughter) the tube house. Two air conditioner units are fitted in the single house, where sensors 0 and 4 were located, that is, the living room and the master bedroom, respectively. The apartment has three air conditioners, located in the three bedrooms (sensors 1, 3, and 4). Surprisingly, no air conditioner is installed in the main living space.

The occupancy profile for both dwelling is broadly similar (Fig. 5). Full occupancy is seen during from the late evening to the early morning, and the lowest percentages are seen during working time. However, some subtle differences can be noticed. First, although the house shows a 60% of occupancy during the central hours of the day, air conditioners are not in use. Conversely, one third of the cooling power is used in the apartment during nigh time and noon; the AC unit located in bedroom 1 is on for almost half of the day.

3.2 Indoor Thermal Environment

The measurements show that the indoor temperature fluctuates within the range of 30–32 °C in the apartment without the use of air conditioning (Fig. 6) and in the range of 29–31 °C in the house (Fig. 7). Relative humidity is remarkably high in both dwellings, with values in the range of 65–80% (Figs. 8 and 9). Although the average August temperature in Ho Chi Minh City is around 28 °C, it is worth noting that a previous study found that more than a half of the central area had land surface temperatures between 31.7 and 34.6 °C in 2016, mainly due to the increased urban density and impervious surface [10]. In other words, these homes are performing quite well in an urban context where the urban heat island effect contributes to temperature increases of about 3–6 °C. It is also noteworthy that the use of air conditioning in the apartment does not lower the indoor temperature that much: DBT drops to 25.5 °C in the conditioned room and to 28–29 °C in the adjacent living room (perhaps because the door connecting the two rooms was left open).

This data would suggest, according to the classic PMV model and the more recent adaptive comfort model, that these houses are rather uncomfortable. Using the online available CBE thermal comfort tool [11] and fixing a few parameters allows for a deeper discussion in this matter. We can assume that the metabolic rate is 1 MET (seated, quiet) and the clothing level is 0.36 CLO (light summer clothes). The air speed, which is necessary to calculate comfort levels, is set at the lowest possible: 0 m/

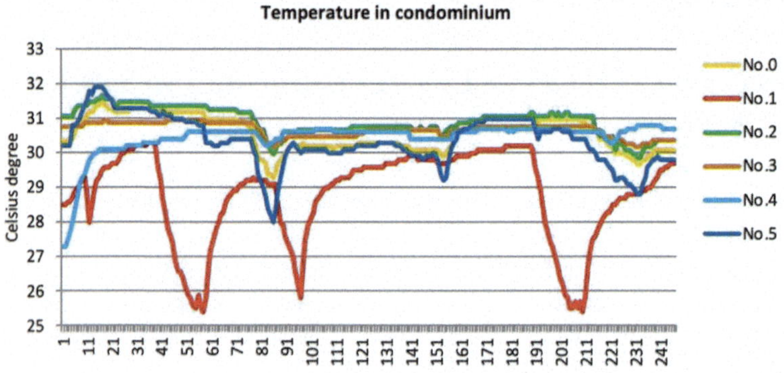

Fig. 6 Temperatures in five different locations inside a condominium unit over six hours

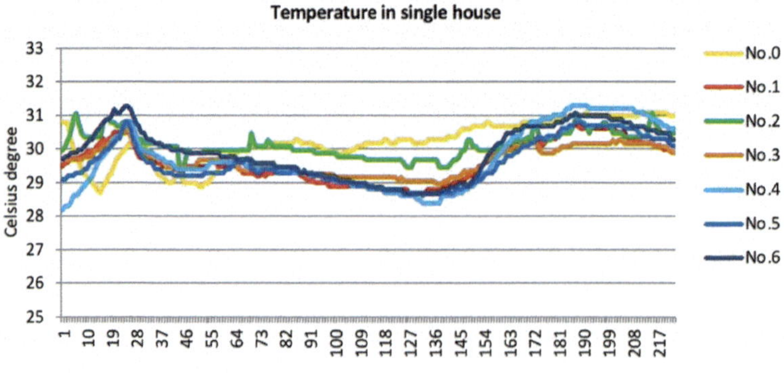

Fig. 7 Temperature in six different locations inside a row house over five and a half hours

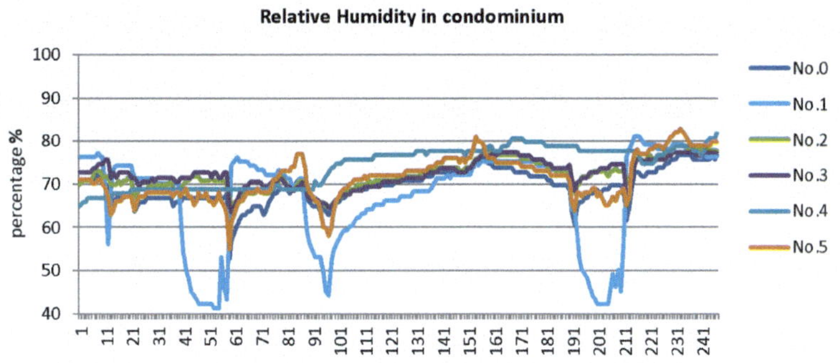

Fig. 8 Relative humidity in five different locations inside a condominium unit over six hours

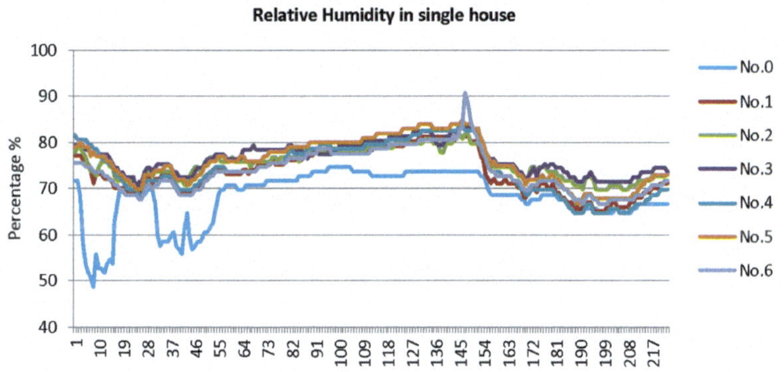

Fig. 9 Relative humidity in six different locations inside a row house over five and a half hours

s for the PMV model, 0.3 m/s for the ASHRAE-55 standard, and lower than 0.6 m/s for the EN-16798. Air temperature and mean radiant temperature are assumed to be the same as the measured DBT, and the outdoor running mean outdoor temperature, when necessary for the calculations, is assumed to be 28 °C.

Three representative conditions of the indoor thermal comfort were selected for evaluation: 32 °C/75% as the upper limit for TBD and RH without air conditioning; 29 °C/75% as the lower limit for TBD and RH without air conditioning; 25 °C/45% as the average conditions for an air-conditioned room (Table 1). The results show that air conditioning is a must to achieve comfort conditions according to the PMV model, which was expected. More interesting discussions arise when considering the two adaptive models. The upper limit of the indoor temperature (32 °C) puts these houses outside the comfort criteria of both standards, but lowering the DBT to 29 °C and the RH to 75% would place them within the 80% range of adaptability (ASHRAE-55) and Class III buildings (EN-16798), provided that the air speed is between 0.3 and 0.6 m/s. It is clear that ventilation plays a fundamental role in these houses, which will be discussed in the next section.

Table 1 Comfort levels according to the ASHRAE-55 and EN-16798 standard

Indoor conditions	ASHRAE-55		EN-16798	
	PMV	Adaptive	PMV	Adaptive
32 °C/75%	2.56 (Cat IV)	Too warm	2.56 (Cat IV)	Class III
29 °C/75%	1.26 (Cat IV)	80% acceptability	1.26 (Cat IV)	Class I
25 °C/45%	– 0.55 (Cat III)	80% acceptability	– 0.55 (Cat III)	Class I (too cool)

With the exception of air-conditioned rooms, indoor conditions border on the warmer zone of thermal discomfort, yet air conditioning is rarely used. Although there is not enough data to draw a definitive conclusion, mainly because the use of air conditioners depends on a variety of factors, including also energy prices and household income, it would be reasonable to assume that cultural norms and thermal adaptation may play an important role. Previous studies have clarified that in low-income households living in cold climates, 80% of residents can express thermal satisfaction at indoor temperature as low as 13.75 °C [12]. Furthermore, in warm tropical climates, adopting setpoint temperatures consistent with a locally developed adaptive comfort model may lead to substantial energy savings in buildings, both for current and future scenarios that take into account climate change [13]. Our results, although limited, suggest that the same may be true in the Vietnamese context, where the development of a local comfort model may prevent the residential sector from falling into the vicious cycle of an increasing energy demand as the results of the introduction of new residential typologies.

Although research on our context, Vietnam, is rather scarce, our results are consistent with studies on the adaptive thermal comfort in tropical locations in East Asia. A meta-analysis of several field surveys in Singapore, Indonesia, Malaysia, China, and Thailand revealed that the most frequent operative temperature in naturally ventilated buildings is about 30 °C and the most frequent relative humidity, between 70 and 80%; furthermore, and it suggests that the operative neutral temperature would be as high as 27.9 °C [14]. Similar conclusions were reached in another study where 97.5% of the 40 surveyed residents in Hanoi expressed thermal comfort at an indoor temperature of 28 °C [15]. The previous literature, together with our results, may provide an explanation of the low use of air conditioners in such warm conditions, as observed in our monitoring.

3.3 Measures for Thermal Comfort

The survey covered several aspects of the residents' living style, but this section presents the most relevant for this chapter, which are related to the passive strategies to regulate the solar radiation, and the use of active devices for climate control. First, a brief outline of the survey demographics is also included to provide with a clear background of the surveyed sample.

127 people responded to the survey, of which 94% lived in Ho Chi Minh City, and 91% said they were between 20 and 40 years old. Therefore, we are talking about a sample of young urbanites. Around three-quarters were living in a row house, 16% in a condominium, and the rest in a detached villa or another types of housing. 89% of the respondents were either students (53%) or company employees (36%).

In terms of home size and occupation, the results were mixed. Most of the houses (23%) were larger than 90 sqm, but smaller houses were not uncommon. Houses between 41 and 50 sqm accounted for 19% of the total, and between 51 and 60 sqm, for 15%. 30% of the properties surveyed housed 4 people, but there were also properties with three (17%), five (18%), and more than five residents (21%). As a result, no clear occupancy profile could be described for our sample.

When asked about the different solar control devices, it is noteworthy that the absence of solar control devices was quite common for all orientations. Curtains on the inside of the windows were the most common strategy in all orientations, followed by overhangs or balconies. Vertical fins, adjustable awnings, and trees were found only in certain orientations, but it is remarkable that all of them are used in east-facing facades (Fig. 10).

The most common active device for climate control was the fan (73%), followed by the air conditioner (21%); 6% of the respondents said that there were no active devices in their homes, and that they only used natural ventilation. Since the number of occupants and the size of the homes were quite spread out, it is natural to assume that the number of air conditioners would also be spread out. Of those who answered that they had air conditioners, 33% said that they had one, 35% said they had two, 21% said they had three, and 11% said they had four or more. What was surprising, however, was to see that 31% of respondents did not use them on a daily basis. Consistent with our monitoring results (Fig. 5), most of the respondents who use air conditioners do so during nighttime (from 21:00 to 6:00) and also around midday

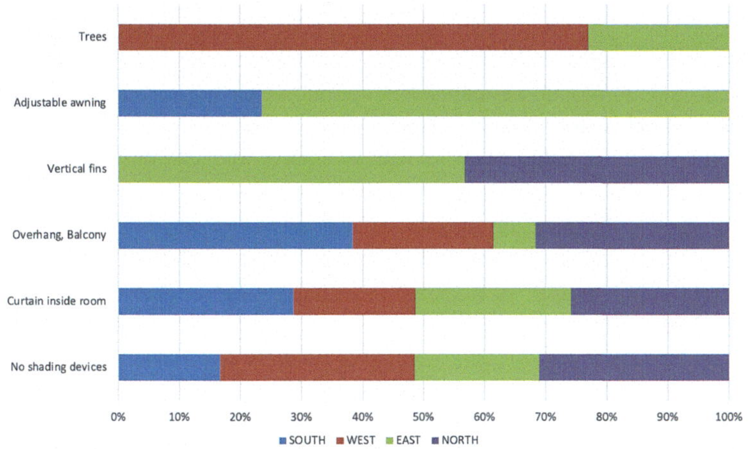

Fig. 10 Shading devices to protect against solar radiation

(from 11:00 to 14:00). We assume that some residents may return home for lunch and take a rest before returning to their duties, but this an assumption that could not be clarified by the actual survey.

4 Main Conclusions and Future Directions of Research

The chapter set out to better understand the perception of thermal comfort and the strategies to achieve it in Ho Chi Minh City, in Vietnam. The monitoring results and the survey to residents in different representative housing typologies have confirmed, in line with similar research, that the comfort conditions in tropical countries deserve detailed examination because, as suggested in our introduction, they may strongly depend on cultural norms and socioeconomic backgrounds.

Our results, although limited, have provided some insight into the comfortable conditions for these residents, with operative neutral temperatures around 28 °C, and relative humidity around 75%. Furthermore, active climate control devices, although present, are not extensively used, suggesting that natural or mechanical ventilation may be sufficient to achieve acceptable levels of comfort. Overall, these suggest that it may be possible to improve comfort conditions while decoupling from the increasing energy demand associated to the extensive use of air conditioning in a country where the air conditioning industry is growing rapidly; from 2013 to 2018, just before the COVID pandemic, sales of room air conditioners in Vietnam grew by 100%, a remarkably higher figure than in other neighboring countries, such as Thailand (18%), Indonesia (4%), or Cambodia (51%) [16].

A number of limitations need to be noted regarding the present study. First, as mentioned above, monitoring was carried out in only two housing units, and in the future this process should be reproduced in more houses to validate our results. Another limitation was that our survey was not standardized, and thus, our results cannot be easily compared with other studies. It is suggested that, in the future, a Post Occupancy Evaluation (POE) survey, either standardized or slightly adapted, should be used to reflect the residents' perception of their houses and their strategies to regulate the indoor thermal environment.

Despite these limitations, this work offers valuable insights for an emerging area in the field of adaptive thermal comfort, which seeks to develop comfort models adapted to local conditions and specific cultural backgrounds. Our findings support recommendations for local governments to develop their own building comfort codes that will allow them to balance the need to improve living conditions with the rational use of energy resources.

References

1. Guy S, Farmer G (2001) Reinterpreting sustainable architecture: the place of technology. J Archit Educ 54:140–148. https://doi.org/10.1162/10464880152632451
2. Nicol JF, Roaf S (2017) Rethinking thermal comfort. Build Res Inform 45:711–716. https://doi.org/10.1080/09613218.2017.1301698
3. World Bank Open Data, World Bank Open Data (n.d.). https://data.worldbank.org (accessed January 5, 2024)
4. Population and housing census in 2019 Migration and urbanization in Vietnam (2019) https://vietnam.unfpa.org/sites/default/files/pub-pdf/migration_and_urbanization_factsheet_eng_final_0.pdf
5. Truong-Young H, Tube Housing (2019) An urban identity of Saigon-Ho Chi Minh City (Doctoral dissertation, La Trobe University). https://s3-ap-southeast-2.amazonaws.com/figshare-production-eu-latrobe-storage9079-ap-southeast-2/38777871/43513_SOURCE01_3_A.pdf?X-Amz-Algorithm=AWS4-HMAC-SHA256&X-Amz-Credential=AKIARRFKZQ25KW2DIYRU/20240105/ap-southeast-2/s3/aws4_request&X-Amz-Date=20240105T004122Z&X-Amz-Expires=10&X-Amz-SignedHeaders=host&X-Amz-Signature=19a66edacc967faf74029ef5a822a2c5086ca97e059e084a78be6949134c789a
6. Pham P, Pham N, Oh S (2021) Spatial configuration of traditional houses and apartment unit plans in Ho Chi Minh City, Vietnam: a comparative study. Spatium. https://doi.org/10.2298/SPAT2145034P
7. Khang HV (2016) Research on architectural design strategies for energy efficiency in residential buildings in Vietnam (Master thesis), The University of Shiga Prefecture
8. Balderstone S, Logan W (2003) Vietnamese dwellings: tradition, resilience, and change, Deakin University, https://dro.deakin.edu.au/articles/chapter/Vietnamese_dwellings_tradition_resilience_and_change/20527458/1 (accessed January 6, 2024)
9. Huynh D (2015) Phu my hung new urban development in Ho Chi Minh City: only a partial success of a broader landscape. Int J Sustain Built Environ 4:125–135. https://doi.org/10.1016/j.ijsbe.2015.03.005
10. Son N-T, Chen C-F, Chen C-R, Thanh B-X, Vuong T-H (2017) Assessment of urbanization and urban heat islands in Ho Chi Minh City, Vietnam using Landsat data. Sustain Cities Soc 30:150–161. https://doi.org/10.1016/j.scs.2017.01.009
11. CBE Thermal Comfort Tool, (n.d.). https://comfort.cbe.berkeley.edu (accessed January 7, 2024)
12. Pérez-Fargallo A, Pulido-Arcas JA, Rubio-Bellido C, Trebilcock M, Piderit B, Attia S (2018) Development of a new adaptive comfort model for low income housing in the central-south of Chile. Energy Build 178:94–106. https://doi.org/10.1016/j.enbuild.2018.08.030
13. Sánchez-García D, Bienvenido-Huertas D, Rubio-Bellido C, Rupp RF (2023) Assessing the energy saving potential of using adaptive setpoint temperatures: the case study of a regional adaptive comfort model for Brazil in both the present and the future. Build Simul. https://doi.org/10.1007/s12273-023-1084-3
14. Nguyen AT, Singh MK, Reiter S (2012) An adaptive thermal comfort model for hot humid South-East Asia. Build Environ 56:291–300. https://doi.org/10.1016/j.buildenv.2012.03.021
15. Nguyen M, Doan V, Pham T, Nguyen V, Nguyen B, Oishi T, Tokura H (2003) Thermal comfort zones in the Vietnamese. J Human Ergol 32:107–110
16. World Air Conditioner Demand by Region (2019) Japan Refrigeration and Air Conditioning Industry Association

Assessing Indoor Hygrothermal Conditions in Unfavourable Scenarios: Open Windows in a Residential Building in the Coastal Mediterranean Summer Climate

Daniel Antón, Amin Al-Habaibeh, Manuel J. Carretero Ayuso, and Madelyn Marrero

Abstract Scientific research has identified different adaptive behaviours that influence thermal comfort in buildings. Depending on factors such as ventilation mode, building type, or occupant age, those user actions are diverse and include air conditioning and fan usage, window and/or door opening, curtain and/or blind usage, cloth adjustment, and beverage drinking, among many others. Especially in the Mediterranean summer climate, façade openings are likely to be a direct inflow of hot air. Focusing on the case study of a residential building with a robust envelope, this chapter aims to assess the impact of keeping windows open on indoor temperature and relative humidity.

1 Introduction

Scientific research has identified different adaptive behaviours that influence thermal comfort in buildings. Depending on factors such as ventilation mode, building type, or occupant age, those user actions are diverse and include air conditioning and fan

D. Antón (✉) · A. Al-Habaibeh
Departamento de Expresión Gráfica E Ingeniería en La Edificación, Escuela Técnica Superior de Ingeniería de Edificación, Universidad de Sevilla, 41012 Seville, Spain
e-mail: danton@us.es

Product Innovation Centre & The Creative and Virtual Technologies Research Lab, School of Architecture, Design and the Built Environment, Nottingham Trent University, Nottingham NG1 4FQ, UK

M. J. Carretero Ayuso
Department of Architectural Graphic Expression, Higher Polytechnic School, University of Extremadura, 10003 Cáceres, Spain

M. Marrero
Departamento de Construcciones Arquitectónicas II, Escuela Técnica Superior de Ingeniería de Edificación, Universidad de Sevilla, 41012 Seville, Spain

M. Torres González and C. Rubio Bellido (eds.), *Towards Low and Positive Energy Buildings*, Green Energy and Technology, https://doi.org/10.1007/978-3-031-70851-0_7

usage, window and/or door opening, curtain and/or blind usage, cloth adjustment, and beverage drinking, among many others [1].

Despite the multifaceted nature of adaptive strategies, research has predominantly concentrated on a limited set of passive measures, such as optimising insulation layers, utilising naturally ventilated envelopes, and employing sun shading devices [2]. Often driven by considerations of energy efficiency and financial implications, these adaptive strategies point out the need for a more comprehensive exploration of alternative approaches.

Chvatal et al. [3] delved into the impact of increased envelope insulation on overall building performance, revealing its beneficial effects during winter but raising concerns about potential overheating in summer. Given the conventional perception of highly insulated envelopes as unequivocal energy-saving solutions, it is therefore necessary to determine optimal insulation levels to avoid both indoor hygrothermal discomfort and increased reliance on air conditioning.

Understanding the influence of outdoor environmental parameters on window-opening behaviour, Kim et al. [4] provided valuable insights for informed design decisions in architecture. According to Mochida et al. [5], cross-ventilation emerges as a key factor and has the potential to significantly enhance indoor thermal comfort, especially in the summer, through strategic control of window openings and tree arrangements. Likewise, Daghigh et al. [6] drew attention to the impact of natural ventilation on indoor humidity and its impact on comfort. Additionally, Tamaskani Esfahankalateh et al. [7] highlighted the significance of window orientation for enhancing thermal comfort in residential buildings, recommending east- and west-facing openings for cold climates. The authors also indicated that heat transfer between the body and the surrounding environment happens through convection, so wind speed turns out to be relevant in this sense.

Li et al. [8] explored the triggers for window opening, emphasising the role of outdoor air temperature and occupants' desire to improve indoor thermal and air quality against outdoor relative humidity, indoor air temperature, indoor relative humidity, and indoor CO_2 concentration.

Further nuances in window-opening patterns were exposed in a year-long study by Lai et al. [9], demonstrating variations across seasons, times of day, and weekdays against weekends. It is worth noting that, in the region of their study (China), opening patterns increased from winter to summer up to around 27 °C, with greater activity during the morning and working hours on weekdays, Window opening on weekends took place to a lesser extent.

The diverse perspectives presented in the literature stress the complexity of adaptive behaviours influencing thermal comfort in buildings. Especially in the Mediterranean summer climate, façade openings are likely to be a direct inflow of hot air and entail a temperature increase, especially in south rooms. According to Zaki et al. [10], occupants may depend on mechanical cooling to counteract the excessive solar gain of rooms, which causes thermal discomfort. In this line, focusing on the case study of a residential building with a robust envelope, this chapter aims to evaluate how keeping windows open affects indoor temperature and relative humidity based on sensor and satellite data collected during a specific two-week period in summertime. Further, this unfavourable scenario was exacerbated by wildfire activity in the immediate vicinity, substantially reducing the area's humidity during that period and for the foreseeable future, thus making it necessary to picture the relative humidity of the case study area. After all, in addition to indoor hygrothermal conditions, thermal comfort is affected by people's adaptation to climate, of which the role of relative humidity is more complex [11].

Having introduced the research context, this chapter is structured as follows: Section 2 describes the case study and its geographical context. Meanwhile, Sect. 3 presents the research methodology, which is supported by own and third-party data on the case study building and its surroundings. Section 4 presents the data achieved and discusses the findings. Finally, Sect. 5 gathers the drawn conclusions and future work.

2 Case Study: A Residential Building with a Robust Envelope

2.1 Context

As seen in Figs. 1 and 2, the case study is located in Alhaurín de la Torre, the seventh most populated municipality in the near vicinity of Málaga city (42,531 citizens in 2022, according to data by the regional government's Instituto de Estadística y Cartografía de Andalucía in 2023 [12]). Geographically speaking, the case study is on the north side of Sierra de Mijas and close to the Guadalhorce River valley and coastal villages. The building is specifically located in Pinos de Alhaurín, a leafy suburb, at approximately 245 m above sea level, near a calcite quarry and a pine forest.

The Sierra de Mijas is a mountainous formation with a variety of flora and fauna, highlighting the presence of the Jarapalos Arboretum, a botanical garden with a multitude of plant species. Unfortunately, these mountains are prone to wildfires (mainly intentioned), especially in summer. In this sense, from 12.30 pm on the 15–17th (stabilised at 7.30 pm) of July 2022, the Sierra de Mijas experienced a devastating wildfire affecting biodiversity. It was not until 19th July that the fire was controlled and finally declared extinguished on 3rd August 2022. It is also worth mentioning that authorities evacuated more than 1500 people preventively since the fire reached, and was likely to reach, populated areas. In relation to the extent (in hectares burnt) of this natural disaster, official numbers differ depending on the reporting public body or organisation: 1788.33 ha [14], approximately 1870 ha by the regional government [15], or 1814.05 ha measured on cartography and 2097.21 ha calculated on the Digital Terrain Model (DTM) of the area a few hours before the fire was stabilised in the evening [16]. Figure 3 shows, in red colour, the surface boundary affected by the wildfire, according to Sentinel 2B satellite data at 10.56 am on 17th July.

Fig. 1 Case study's geographical context: Sierra de Mijas and coastal surroundings. Viewed from the west. *Source* based on Google Earth Pro 3D content [13]

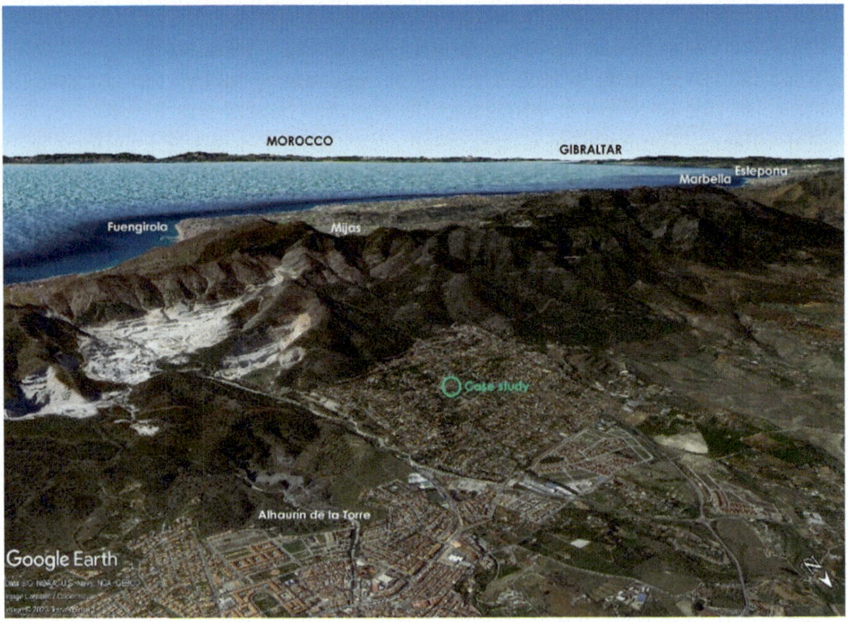

Fig. 2 Case study's geographical context: Sierra de Mijas and coastal surroundings. Viewed from the northeast. *Source* based on Google Earth Pro 3D content [13]

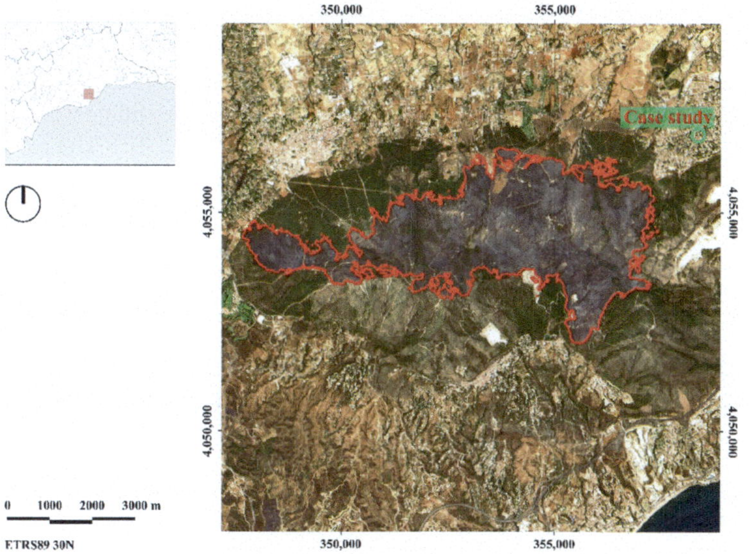

Fig. 3 Location and extent of Sierra de Mijas' wildfire on 17th July 2022. *Source* based on Javier Martos' map [16]

Section 4 will reveal the role of this foliage-covered sierra in the area's moisture, but now that it has been almost destroyed, the climatic implications of this wildfire are yet to be determined.

2.2 Building Typology and Features

The construction of the case study building concluded in 1999, several years before the entry into force of the continuously amended and current Technical Building Code [17], the reference set of regulations that establish the requirements and standards for building design and construction in Spain. Notwithstanding, despite the regulations applicable to date (1970s–1980s), the building was provided with a robust envelope with thermal insulation and a significant mass in walls, Thus yielding a reasonable theoretical thermal inertia.

The case study's building was a ground-floor detached house with a reinforced concrete structure based on pillars and slabs and a sloping roof. The walls forming the thermal envelope were 34.5 cm thick and consisted of the following layers, from the outside to the inside: First, white vibrated concrete rough blocks (there was no additional cladding), A 2 cm layer of cement mortar to seal the blocks and receive the projection of 2 cm of polyurethane foam. Next, Letting a non-ventilated 3 cm cavity in the inner core of the wall to reduce thermal transmittance, A hollow-brick wall with a thickness of 7 cm served as the basis of the 1.5 cm thick plastering in contact with indoor ambient.

The two monitored rooms of the building had different orientations at opposite façades, Particularly south and north. Each room comprised one sliding window with an effective opening surface of approximately 0.50 m^2 (0.45 × 1.10 m) for each open sash. These windows allowed cross-ventilation throughout a corridor.

3 Methodology

The research methodology comprises two different data sources: (1) Satellite imagery and data, and (2) Data recorded in situ from the case study building.

3.1 Satellite Data Collection

3.1.1 Satellite Imagery

Sentinel-2 L2A data were used to contextualise the case study in terms of geomorphological features, its proximity to the sea, and the surrounding dryness in order to supplement recorded hygrothermal data from the case study building.

Access to satellite data was granted via the Copernicus Browser from the Copernicus Data Space Ecosystem [18]. According to this online system, Sentinel-2 satellite provides high-resolution visible and infrared spectral band images to monitor vegetation, soil and water cover, inland waterways, and coastal areas to achieve land-cover maps, land-change detection maps, vegetation monitoring, or monitoring of burnt areas. Level 2A data, available since June 2015 and March 2017 globally, are high-quality data processed to filter atmospheric effects on the light reflected off the Earth's surface, reaching the sensor.

Collected satellite imagery comprised two distinct types:

- 'True-colour' images (visible light) consisting of the red, green, and blue (RGB) colour channels to produce a natural coloured representation of the Earth, as seen with the naked eye.
- 'Normalised Difference Moisture Index' (NDMI) images, with values ranging from − 1 to 1 (coefficient). Low, negative values (intense red) indicate barren soil, values around zero (from approximately − 0.2 to circa 0.4) generally correspond to water stress, and high, positive values (dark blue) account for high canopy with no water stress. In this way, NDMI images are used to determine vegetation water content and monitor droughts.

The time span of satellite image recording was determined by the following factors: (i) the hygrothermal monitoring period of the case study building, so that it was mostly covered (from 9th to 23rd July 2022), (ii) the wildfire that took place nearby during that period, and (iii) the comprehensiveness and quality of satellite imagery. The second factor made it convenient to explore the area's moisture conditions before and after the natural disaster, which happened from noon on the 15th until the evening of the 17th of July 2022, in order to reveal a possible impact on the indoor and outdoor temperature and relative humidity values of the case study. The latter factor, i.e. the suitability of image data, was affected by the coverage of satellite flight paths over the area and its cloudiness. In an effort to reach the end of the building's monitoring period, satellite imagery was consulted in the days that followed. However, the images were incomplete. It was not until 1st August that the whole area was captured by Sentinel-2 (the moisture conditions may have undergone alterations over that extended time frame) except for 27th July, but the image presented a significant number of clouds covering the Sierra de Mijas. Consequently, it was decided to display 22nd July as the closing instance of the satellite observation. Similarly, pre-wildfire data were retrieved from the Copernicus Browser dated 27th June for comparison, thus ensuring the completeness mentioned above and the quality of image data.

Beyond collecting the satellite images (True colour and NDMI), pre—and post-wildfire histograms were downloaded from the NDMI images, so that the moisture index could be more easily visually quantified and understood in terms of significance. In addition, both NDMI images were converted into greyscale to quantify the pixel intensities as per [19] and as seen in [20]. This was carried out by enquiring the digital number (DN) in 8-bit encoding (256 values ranging from 0 to 255). As a consequence, the lost vegetation and the new surface with extreme dryness could be represented by means of two supplementary histograms.

3.1.2 Meteorological Data

This research also benefitted from meteorological satellite data, which were useful in capturing the outdoor conditions of the case study building's surroundings during indoor hygrothermal monitoring. This was achieved via the POWER Data Access Viewer by NASA [21] to obtain higher resolution daily time series data from the Prediction of Worldwide Energy Resource (POWER) Project. With its origin in NASA research and managed by NASA's Global Modelling and Assimilation Office (GMAO), POWER provides solar and meteorological data sets to support renewable energy, energy efficiency in buildings, and agricultural needs. Here, meteorological data are provided by the Modern-Era Retrospective analysis for Research and Applications, Version 2 (MERRA-2) Project, available since 1st January 1981, with data until within several months of real time [22].

In this research, by entering the case study's coordinates (hidden for privacy), a comprehensive series of meteorological data was extracted from MERRA-2:

- The average air (dry bulb) temperature at 2 m above the Earth's surface (°C) indicated as T2m in this chapter's tables.
- Dew point (temperature) at 2 m (°C)

- Relative humidity at 2 m (%), indicated as RH2m in this chapter's tables.
- Surface pressure (kPa)
- Wind speed at 10 m (m/s), WS10 in tables
- Wind direction at 10 m (degrees)
- Wind speed at 50 m (m/s), WS10 in tables
- Wind direction at 50 m (degrees)

Among those parameters, the most relevant were chosen to analyse the building performance considering the season of the year, the case study altitude, and prioritising data deriving from the sensors installed.

Finally, the 2022 mean energy intensity or mean solar radiation on the case study area's Earth surface was also obtained to contextualise the case study.

3.2 *Sensor Data: Indoor and Outdoor Conditions*

Data revealing the case study's indoor and outdoor conditions consisted of temperature and relative humidity recordings. This was carried out using three OMEGA data loggers (OM-CP-RHTEMP101A) attached to the walls and configured to monitor strategic case study locations: a bedroom facing the south façade (ID 1_South), a guest room facing north (ID 2_North), and outside the latter room's window, protected from direct sunlight (ID 3_Exterior). It should be noted that both rooms are opposite, thus allowing cross-ventilation. Considering that the aim of this research was to ascertain the building's hygrothermal performance, no cooling was used during the monitoring period, letting its passive features be the sole influencing factors.

The temperature and relative humidity monitoring took place every 15 min during the 14 days from 9th to 23rd July 2022, commencing and concluding at 12 pm CEST (Central European Summer Time). In the course of that period, records were kept for both instances: (i) when windows in any of the rooms were closed and (ii) the occupancy status of each room.

After acquiring the data, temperature and relative humidity values were concurrently visualised in individual graphs for each sensor. This presentation aimed to facilitate the comprehension of the relationship between the two variables. Additionally, key statistical parameters, including maximum, minimum, mean, and standard deviation were computed. The analysis approach extended to encompass the coefficient of variation, covariance, and correlation coefficient, contributing to a comprehensive characterisation of their variability and interrelationship.

4 Results and Discussion

As per the methodology structure above, this section presents and analyses the data collected by following both qualitative and quantitative statistical approaches.

4.1 Satellite Data

4.1.1 Sentinel-2 L2A Data

In this research, satellite image data are used to depict the case study area's environmental conditions. In this sense, Fig. 4 shows Sentinel-2 L2A true-colour and NDMI images and histograms before and after the wildfire occurred in Sierra de Mijas. The various data reveal a clear difference between both scenarios, highlighting the significant loss of vegetation towards the barren surface, especially with values ranging from − 0.08 to − 0.30 in the NDMI images, also seen as forest burnt in the true-colour image. Consequently, as burnt areas retain less moisture than foliage-covered sectors, an impact on the region's climate is expected. In turn, this could impact the hygrothermal conditions of the case study as opposed to the situation in which the wildfire did not occur.

In addition, although the conversion of RGB images into greyscale depends on colour interpretation and assignment to index values, pixel intensities were enquired to approximate the quantification of vegetation losses further. The following histograms illustrate the lost vegetation and the new surface with extreme dryness. Thus, based on the converted RGB NDMI images, Fig. 5 quantifies the number of pixels corresponding to this index.

Also shown as the percentage of higher presence (Fig. 6):

In view of the above, a lower DN represents vegetation, whereas higher DN values account for dry areas, the latter mainly corresponding to burnt surfaces. Thereby, from DN 117 onwards, it is the post-wildfire scenario that exhibits a higher presence of pixels corresponding to areas with increased water stress, away from the extremely high percentage of pre-wildfire vegetation from DN 0 to circa 50 and the more slight foliage increase within the 50–117 range.

4.1.2 MERRA-2 Meteorological Data

Satellite meteorological data collected from NASA's MERRA-2 Project are presented here to determine the quantities applicable to the case study. To do this, these quantities are displayed together by plotting their values throughout the hygrothermal monitoring period set.

Figure 7 shows MERRA-2 temperature and relative humidity data, which are estimations at 2 m over the Earth's surface. In order to assess their suitability for this

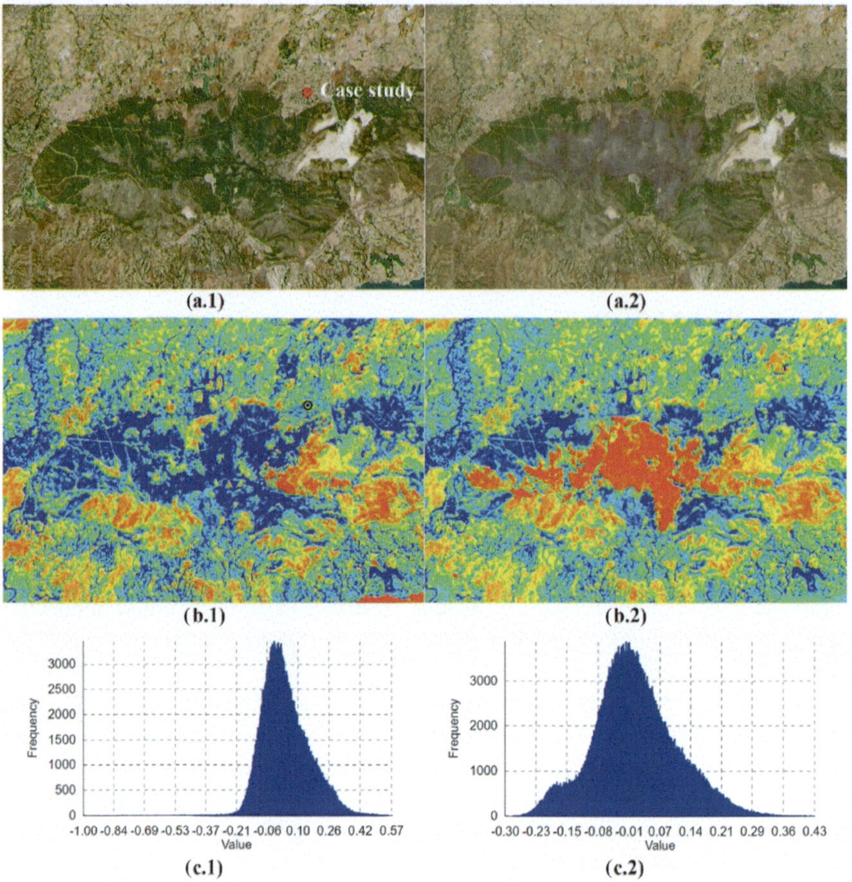

Fig. 4 Sentinel-2 L2A satellite data from Sierra de Mijas' wildfire, pre- (1, left column) and post- (2, right column) wildfire: **a** true colour images; **b** NDMI images; and **c** NDMI histogram. *Source* NASA's 'POWER Project' Data Access Viewer [21]

study, it is therefore necessary to compare them with outdoor data recorded from the case study building (sensor ID 3_Exterior), although the latter will be addressed in the following subsection.

To support that decision, quantitative, statistical data from the two-source meteorological parameters above are gathered in Table 1.

All considered, both the graph and the analysis results reveal significant differences between data from the two sources, which may lead to wrong insights into the case study's hygrothermal behaviour: MERRA-2's average RH is 73% higher than that deriving from Sensor 3 ID 3_Exterior, while the mean temperature from the former is approximately 11% lower during the monitoring period. Moreover, MERRA-2's temperature rises with sunlight before the actual temperature does, the latter being extracted from Sensor 3's log. This fact is qualitatively shown in the graph

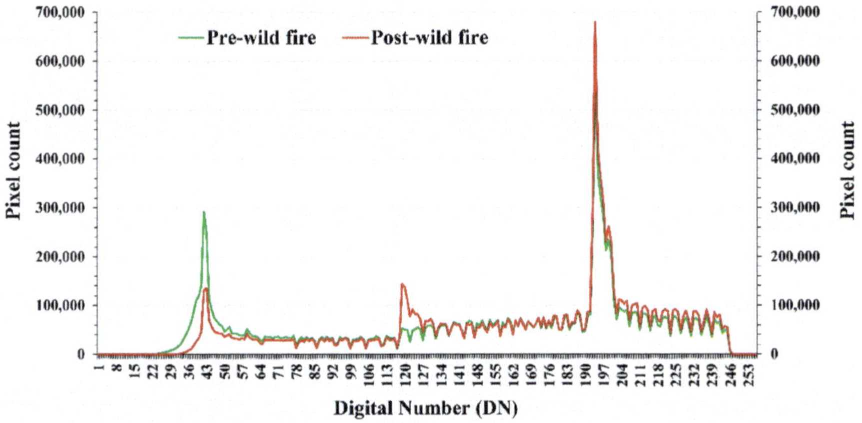

Fig. 5 Histogram of pre- and post-wildfire pixel distribution

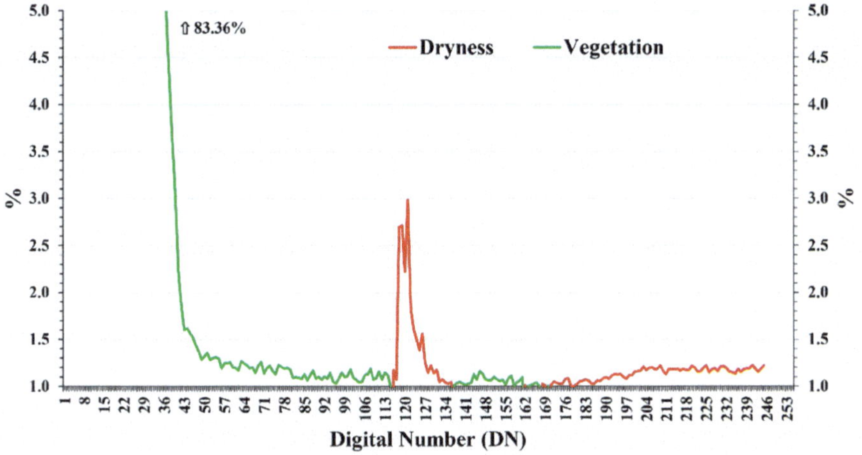

Fig. 6 Histogram of vegetation and dryness pixel distribution

above and from the calculated covariance, with a positive relationship revealing a common trend. It is worth noting that negative values indicate an indirect relationship between variables (this is the case of temperature and relative humidity), whereas positive values show that they jointly increase and decrease. Besides, considering their high covariance, the strong direct relationship between relative humidity from the two sources is remarkable since when the recorded RH (sensor) increases, the estimated RH (satellite) rises at a higher rate, which could be away from the ground-floor reality, the case study.

This research did not aim to understand the difference between data from the cited sources (recorded against estimated), but it may be partly due to the fact that the case

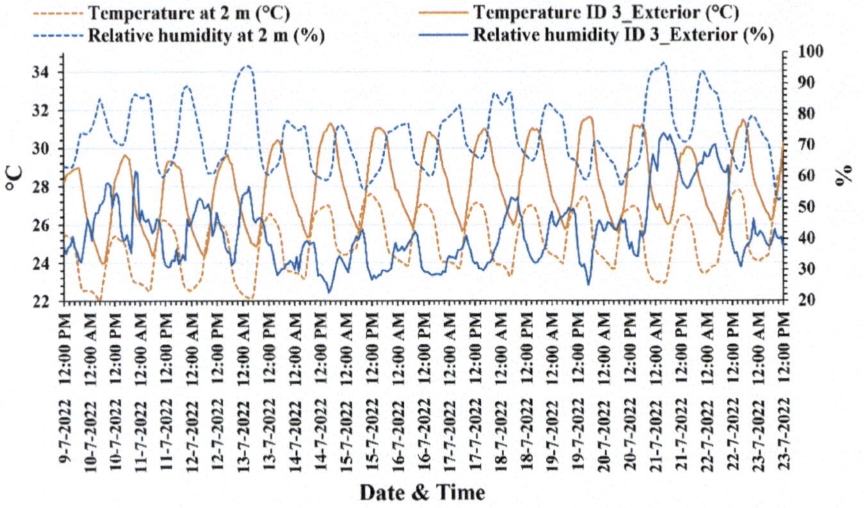

Fig. 7 Case study monitoring: outdoor ambient temperature and relative humidity. *Source* NASA's 'POWER Project' Data Access Viewer and sensor ID 3_Exterior, respectively [21]

Table 1 Statistical analysis of data collected from MERRA-2 and Sensor 3

	Max*	Min*	Mean	SD	CoV*	Covariance	Correlation coef
T2m (°C)	28.83	22.05	25.02	1.56	0.06	1.82	0.61
T Sensor3 (°C)	31.62	24.00	28.05	1.94	0.07		
RH2m (%)	96.56	52.38	72.97	10.35	0.14	61.89	0.54
RH Sensor3 (%)	73.90	22.70	42.18	11.06	0.26		

*Where 'Max' accounts for maximum, 'Min' for minimum, and 'CoV' for coefficient of variation, and 'T' and 'RH', respectively, represent temperature and relative humidity

study presents obstacles to wind, such as the building itself or its surroundings, particularly, a significant number of trees and other nearby constructions.

For all those reasons, it is the temperature and relative humidity recorded by Sensor ID 3_Exterior that is selected to depict the case study's performance.

In addition, Fig. 8 depicts how wind speed at 10 m and 50 m altitude (from the Earth's surface) and surface pressure behaved during the monitoring period.

The wind direction at 10 m is not represented in the graph above but follows a quasi-regular behaviour during the monitoring period. However, it was from 8 am on the 19th of July until 12 am on the 20th that the wind experienced a change in direction, progressively blowing from southeast to northwest during that lapse. Given the representation of wind speed in Fig. 8, the clear drop in its values corresponds

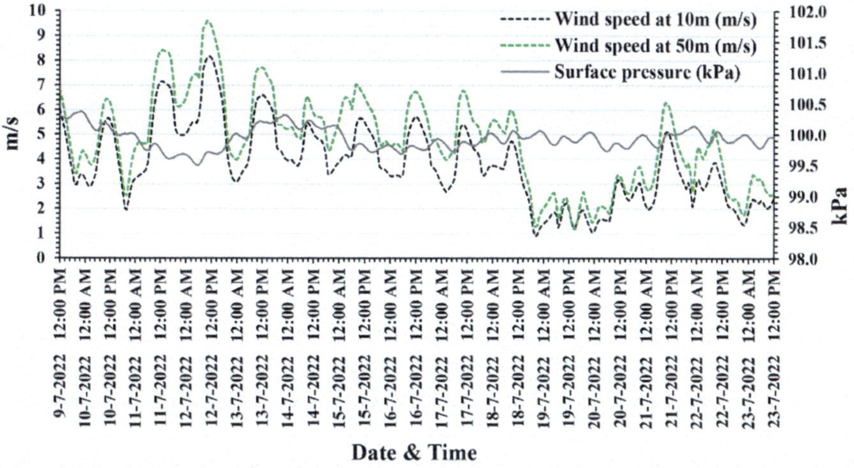

Fig. 8 Case study monitoring: wind speed at 10 m and 50 m, and surface pressure. *Source* NASA's 'POWER Project' Data Access Viewer [21]

to the cited change in wind direction, but there is no noticeable impact on the rest of the quantities, at least qualitatively.

Besides, in view of the lower values of wind speed at 10 m in relation to that at 50 m, a reduction in wind speed can be anticipated at a lower altitude (1–2 m against 10 m). As explained above, near the ground level, the building itself and its surroundings constitute obstacles hindering the wind, so this should also have an impact on temperature and relative humidity estimations by MERRA-2. For its part, the dew point (temperature) at 2 m was not analysed, as the recorded (sensor) outdoor temperature was always higher, even at night.

Regarding surface pressure, Fig. 8 above reveals clearly defined rises and falls in cycles of approximately 12 h each. In light of recorded temperature cycles (of around 24 h), surface pressure also turns out to be a relevant parameter to study. Therefore, wind speed at 10 m and surface pressure were chosen as the key meteorological (MERRA-2) quantities for the joint analysis with hygrothermal monitoring data.

Finally, it is worth providing the 2022 mean energy intensity or mean solar radiation on the case study area's Earth surface: 4.79 kWh/m2/day.

4.2 Case Study Building Temperature and Relative Humidity

Considering recordings every 15 min, 2690 data points were collected from each sensor's log at the end of the monitoring period (1345 values for each quantity, temperature, and relative humidity). This subsection is structured as follows: indoor sensors (two devices in rooms with a different orientation) and a third sensor installed outside the building (north façade).

4.2.1 Sensors 1 and 2: Indoor (South and North)

Here, temperature and relative humidity data from indoor sensors are first plotted and statistically analysed for each device separately. Thus, Fig. 9 illustrates the behaviour of those quantities in the south bedroom (Sensor 1).

Quantitative data from the previous curves are shown below (Table 2):

The data shows what could be considered high-temperature parameters for an indoor space. With windows open most of the monitoring period, the south bedroom presents an average of 2.44 °C daily difference between day and night in temperature (relative maxima and minima). There is also a greater degree of dispersion of relative humidity in contrast to indoor temperature in this sunlit room.

In terms of how the variables behave in relation to each other, covariance and correlation coefficient indicate a strong, indirect relationship between them in such a way that relative humidity markedly decreases as temperature increases, as expected, especially after noon. Relative humidity is unstable and dispersed, with values scattered across each day–night cycle. In contrast, temperature is more predictable, with

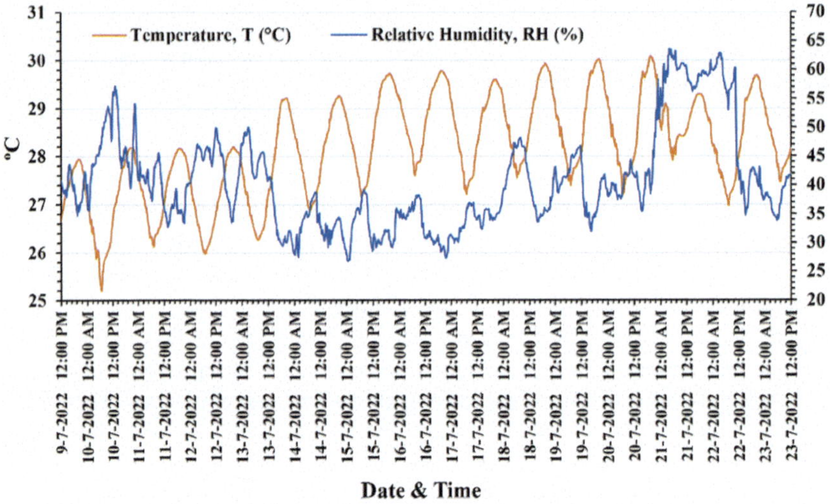

Fig. 9 Case study monitoring: Sensor 1's temperature and relative humidity. South bedroom

Table 2 Statistical analysis of data collected by Sensor 1. South bedroom

	Max	Min	Mean	Standard deviation	CoV	Covariance	Correlation coefficient
T Sensor1 (°C)	30.09	25.21	28.18	1.03	0.04	− 2.53	− 0.29
RH Sensor1 (%)	63.60	26.80	40.75	8.48	0.21		

more regular increases and decreases for each cycle, despite the great ordinate (y-axis) cycle breadth.

On its end, the behaviour of temperature and relative humidity of Sensor 2's data (north guest room) are displayed in Fig. 10.

Table 3 gathers the statistical parameters from the above graph:

In comparison with Sensor 1, the north guest room experienced a more regular temperature as the data were less dispersed, both qualitatively and quantitatively, although further analysis below will show differences between their thermal behaviour. In the north guest room, the average daily temperature difference between day and night was approximately 0.81 °C. About relative humidity, records from Sensor 2 were similar on average but reached lower and higher absolute maximum and minimum values. In addition, the graph and the statistical analysis reveal an overall common behaviour of the temperature and the relative humidity.

Once data from indoor sensors have been separately displayed and analysed, it is worth presenting and studying both sources jointly in order to depict the hygrothermal performance of the case study building depending on the orientation. Prior to this,

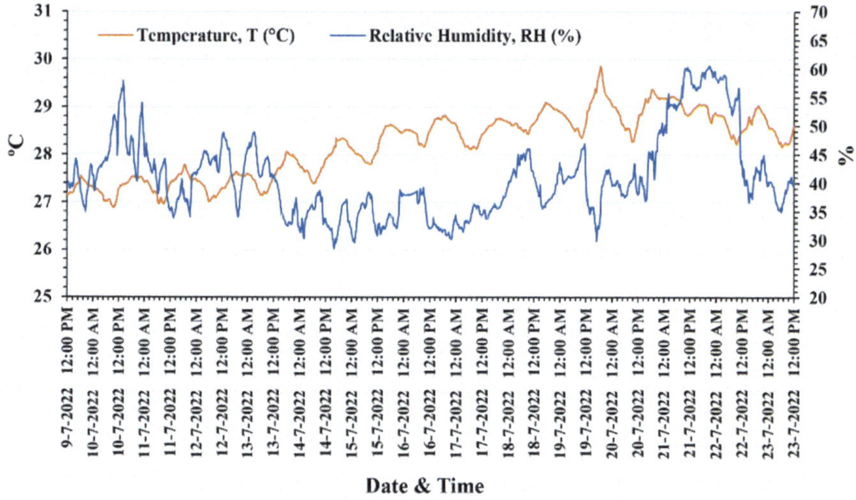

Fig. 10 Case study monitoring: Sensor 2's temperature and relative humidity. North guest room

Table 3 Statistical analysis of data collected by Sensor 2. North guest room

	Max	Min	Mean	Standard deviation	CoV	Covariance	Correlation coefficient
T Sensor2 (°C)	29.86	26.88	28.21	0.68	0.02	0.31	0.06
RH Sensor2 (%)	60.60	28.60	40.95	7.22	0.18		

times when windows were closed because of events such as the wildfire and occupancy periods of the south bedroom are also specified (Tables 4 and 5)—there was no occupation in the north guest room during the monitoring period.

Figure 11 shows closed windows and occupancy along with temperature values from the two sensors installed.

Table 4 Window closing and occupancy in case study rooms

	Date	9th July	10th July	11th July	16th July	17–18th July
Closed windows	Room	–	North		Both	South
	Time	–	1.00–9.45 pm		12.00–6.15 am	3.00–5.40 pm
Occupancy	Time	3.00–5.30 pm	1.00–10.00 am	1.00–10.00 am		11.45 pm–8.00 am
	People	1	2	1		2

Table 5 Occupancy in the north guest room

	Date	18–19th July	20th July	21st July	22nd July	23rd July
Occupancy	Time	11.45 pm–8.00 am	2.00–10.30 am	1.00–10.30 am	1.30–9.30 am	1.00–8.30 am
	People	1	1	1	2	1

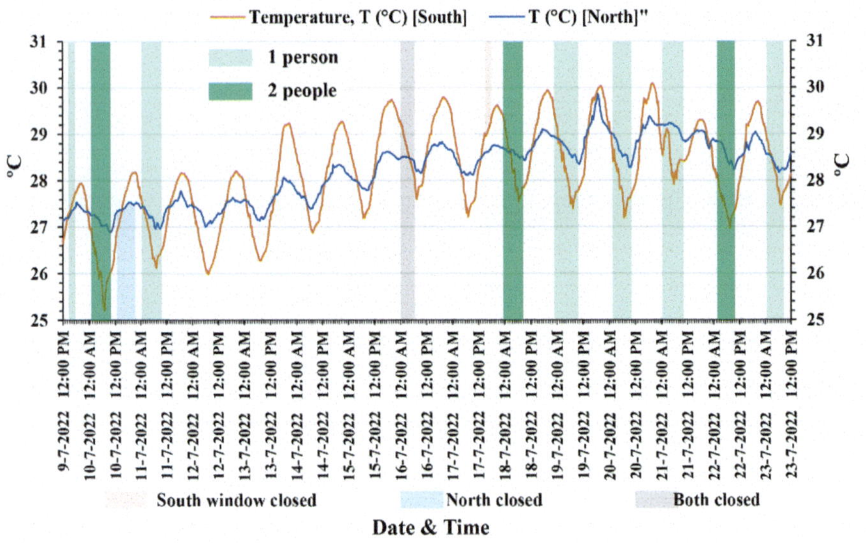

Fig. 11 Case study monitoring: Sensors 1 and 2's temperature. South bedroom and north guest room

From the above, the situation when both windows were closed is also worth discussing. Here, temperatures decreased at night at a slower pace, particularly in the north guest room and more slightly in the south bedroom. Similarly, the routine temperature increases during sunlight followed a slower pace than usual for the two sensors (1 and 2) in cases of window closing.

For its part, Table 6 compiles the breakdown of relative maximum (daytime) and minimum (nighttime) temperature values of daily cycles during the monitoring period. This allows for the study of different room orientations in relation to each other.

Thus, the mean relative maximum temperature recorded by Sensor ID 1_South during the monitoring period is 2.41% higher than that deriving from the north guest room monitoring; likewise, the average relative minimum temperature from the latter sensor (Sensor ID 2_North) is circa 3.40% higher. These facts reveal that the case study building experiences slightly better indoor thermal comfort (lower temperature in summertime) in the north non-sunlit guest room during the day, as expected. In contrast, despite occupancy, the sunlit room (south) is more exposed to changes, thus seeing a more severe temperature decrease during the night, which in turn means better thermal comfort under the studied conditions of the case study. The relative humidity recorded by both sensors was also jointly analysed (Fig. 12).

Table 6 Indoor relative maximum and minimum temperature values of day cycles

Day cycle	Daily T_{max} * (South) (°C)	Daily T_{max} (North) (°C)	T_{max} Increase (*S*/*N*)*	Daily T_{min}* (North) (°C)	Daily T_{min} (South) (°C)	T_{min} Increase (N/S)
09/07/2022	27.95	27.54	1.01	–	–	–
10/07/2022	28.19	27.54	1.02	26.88	25.21	1.07
11/07/2022	28.17	27.79	1.01	26.96	26.14	1.03
12/07/2022	28.21	27.64	1.02	27.01	25.99	1.04
13/07/2022	29.23	28.07	1.04	27.14	26.28	1.03
14/07/2022	29.27	28.34	1.03	27.39	26.88	1.02
15/07/2022	29.74	28.62	1.04	27.80	27.19	1.02
16/07/2022	29.79	28.83	1.03	28.16	27.61	1.02
17/07/2022	29.61	28.77	1.03	28.11	27.22	1.03
18/07/2022	29.94	29.10	1.03	28.43	27.56	1.03
19/07/2022	30.03	29.86	1.01	28.35	27.40	1.03
20/07/2022	30.09	29.38	1.02	28.28	27.20	1.04
21/07/2022	29.31	29.08	1.01	28.83	27.92	1.03
22/07/2022	29.70	29.05	1.02	28.23	26.98	1.05
23/07/2022	–	–	–	28.17	27.47	1.03

* Where 'T_{max}' and 'T_{min}' account for maximum and minimum temperature, respectively, '*S*' means South, and '*N*' is North

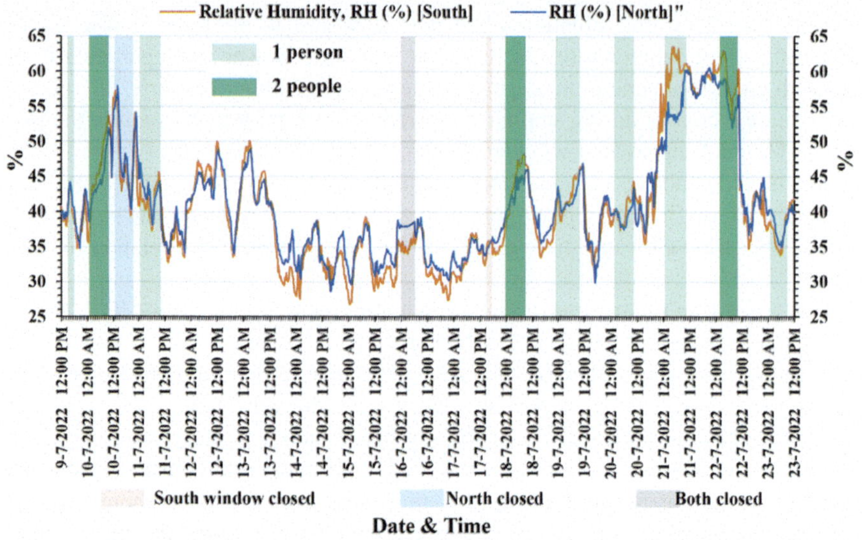

Fig. 12 Case study monitoring: Sensors 1 and 2's relative humidity. South bedroom and north guest room

Given the complexity of relative humidity and the presence of multiple relative maximum and minimum values throughout the daily cycles, they were not calculated. Instead, the relationship between the relative humidity recorded by each sensor was quantified, obtaining a covariance of 59.59 and a correlation coefficient of 0.97.

Also supported by the graph above (Fig. 12), the statistics indicate a strong direct relationship between the data from both sensors. It is worth noting that occupancy makes the south bedroom's relative humidity higher at night since the north guest room is generally superior the rest of the time.

4.2.2 Sensor 3: Outdoor (North Façade)

This subsection includes the analysis of (i) the case study's outdoor ambient conditions, studying data recorded by Sensor ID 3_Exterior and relevant meteorological parameters deriving from MERRA-2, and (ii) the behaviour of Sensor ID's 1_South and 2_North against 3_Exterior in order to characterise the building performance.

Firstly, given that temperature and relative humidity values from the building monitoring were recorded every 15 min, in contrast to the hourly recording for wind speed at 10 m (MERRA-2 data), values corresponding to the 15, 30, and 45 min between hours were linearly interpolated to complete the satellite data. As a result, the joint analysis of the variables is straightforward, with a wind speed value for every temperature and relative humidity datum (Fig. 13).

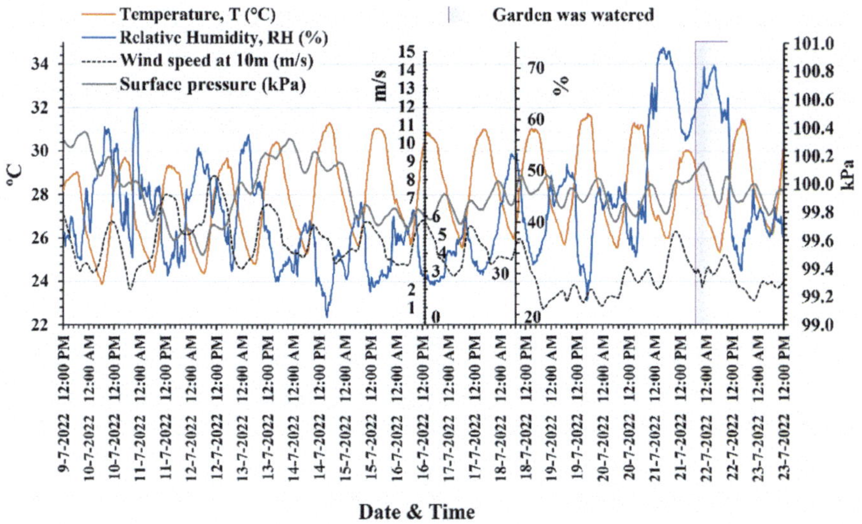

Fig. 13 Case study monitoring: outdoor ambient temperature and relative humidity, wind speed and surface pressure. *Source* sensor ID 3_Exterior and NASA's 'POWER Project' Data Access Viewer, respectively [21]

As previously seen, Table 7 gathers statistical parameters to contrast quantities recorded by a sensor installed in the case study building and, this time, satellite meteorological data.

The surface pressure remains predominantly stable throughout the monitoring period. In view of Fig. 13, the 12-h cycles of surface pressure correspond to fluctuations in relative humidity, whereas temperature exhibits cycles of approximately 24 h. Surface pressure rises and falls rapidly alongside temperature, coinciding with the onset and conclusion of each temperature cycle. However, two cycles of surface pressure occur during the solar radiation cycle, at approximately the highest temperature value, making the surface pressure reach a lower value.

Regarding the behaviour of quantities between each other, it is worth highlighting the accordance of outdoor temperature with wind speed and surface pressure. In contrast, relative humidity is significantly influenced, especially by wind speed, to a greater extent. It has an indirect relationship with wind speed and a direct, albeit lesser, connection with surface pressure.

Secondly, outdoor sensor data were compared to those from Sensors 1_South and 2_North, so that the building's hygrothermal behaviour could be depicted (Fig. 14).

In the graph above, the vertical scale remained constant to facilitate the identification of extreme values from each of the three temperature data sources.

For its part, the relationship between indoor and outdoor temperature values from each sensor was quantified in Table 8.

Table 7 Statistical analysis of sensor temperature and relative humidity data against meteorological data. [21] Source: own data and NASA's 'POWER Project' Data Access Viewer, respectively

	Max	Min	Mean	Standard deviation	CoV	Covariance	Correlation coefficient
T Sensor3 (°C)	31.62	24.00	28.05	1.94	0.07	– 11.30	– 0.53
RH Sensor3 (%)	73.90	22.70	42.18	11.06	0.26		
Surface pressure (kPa)	100.32	99.50	99.92	0.16	0.00		
				SurPress* against T sensor3	– 0.03	– 0.10	
				SurPress against RH sensor3	0.40	0.22	
				SurPress against WS10*	0.01	0.04	
Wind speed at 10 m (m/s)	8.16	0.89	3.80	1.57	0.41		
				WS10 against T sensor3	0.00	0.00	
				WS10 against RH sensor3	– 2.40	– 0.14	

*Where 'SurPress' accounts for surface pressure and 'WS10' for wind speed at 10 m

Table 9 indicates a more direct and stronger relationship between outdoor temperature and that from the south bedroom, which is more extreme than the temperature from Sensor 2.

In addition, relative maximum and minimum temperature values of daily cycles were also calculated (Table 9); this time, the most unfavourable scenarios were addressed, such as the maximum indoor temperature values, both daytime and night-time, as they affect thermal comfort. The maximum temperature of the 23rd of July and the minimum of the 9th were not calculated as those days were not fully monitored.

Given the results above, the average relative maximum outdoor daytime temperature is 4.6% higher than that recorded by Sensor 1 in the south bedroom, which, in turn, represents a mean difference of approximately 1.36 °C. In contrast, the non-sunlit north guest room shows an average relative minimum indoor temperature of around 10% higher than the exterior at night, reaching a mean difference of 2.55 °C, which, by itself, may entail thermal discomfort inside the building under those conditions.

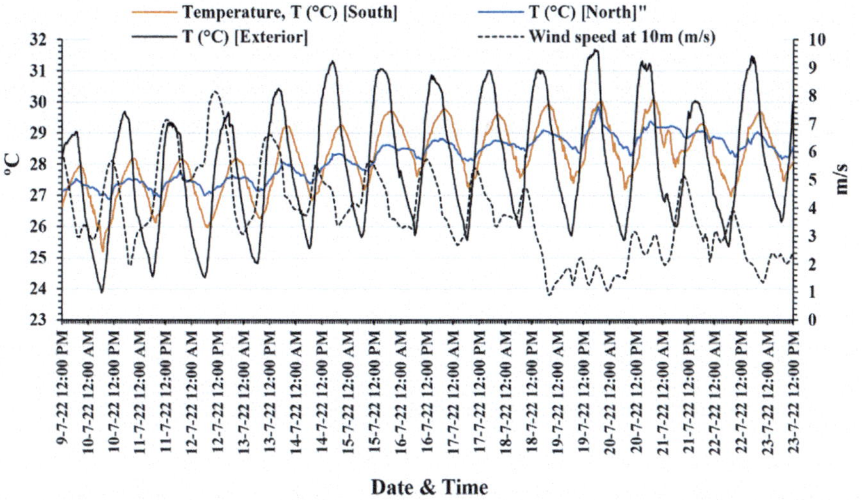

Fig. 14 Case study monitoring: indoor and outdoor ambient temperature and wind speed. *Source* sensors 1, 2, and 3, and NASA's 'POWER Project' Data Access Viewer, respectively [21]

Table 8 Statistical analysis of temperature data collected by all sensors

	T Sensor 3 versus 1	T Sensor 3 versus 2
Covariance	1.30	0.58
Correlation coefficient	0.65	0.44

Besides, the typical temperature drop in the late afternoon, particularly in the south-facing room—it is the most exposed to solar radiation—occurs with a delay of approximately 2–4 h compared to the exterior. This may be due to the thermal inertia effects of the building envelope still applying during this period, as it absorbs energy throughout the day. In contrast, due to the thermal envelope's characteristics and as a result of solar activity since sunrise, the temperature increase in the building takes between 1 and 2 h in comparison with the outside.

Regarding the wind, Table 10 gathers statistical parameters of its relationship with sensor temperature data.

These data show the opposite behaviour of wind against indoor temperatures, as they generally increase when wind speed decreases. What is more, the unstable, nonuniform curves of temperatures correspond to periods with irregular wind speeds, with drops and rises of a different ordinate (y-axis) cycle breadth. This could indicate that wind speed affects indoor temperature behaviour and partly the outdoor temperature while sharing an overall common trend with the latter.

With respect to relative humidity, this research revealed its complexity both inside and outside the building. For the first time in this chapter, relative humidity data from sensors 1, 2, and 3 are displayed together, so that their complex behaviour is depicted (Fig. 15). Again, the scale of the vertical axis remains constant.

Table 9 Indoor and outdoor relative maximum and minimum temperature values of day cycles

Day cycle	Daily T_{max} (Exterior) (°C)	Daily T_{max} (South) (°C)	T_{max} Increase (Ext/S)	Daily T_{min} (North) (°C)	Daily T_{min} (Exterior) (°C)	T_{min} Increase (N/Ext)
09/07/2022	29.04	27.95	1.04	–	–	–
10/07/2022	29.70	28.19	1.05	26.88	23.88	1.13
11/07/2022	29.35	28.17	1.04	26.96	24.41	1.10
12/07/2022	29.67	28.21	1.05	27.01	24.37	1.11
13/07/2022	30.44	29.23	1.04	27.14	24.82	1.09
14/07/2022	31.32	29.27	1.07	27.39	25.32	1.08
15/07/2022	31.06	29.74	1.04	27.80	25.67	1.08
16/07/2022	30.87	29.79	1.04	28.16	25.73	1.09
17/07/2022	31.01	29.61	1.05	28.11	25.58	1.10
18/07/2022	31.04	29.94	1.04	28.43	25.97	1.09
19/07/2022	31.72	30.03	1.06	28.35	25.71	1.10
20/07/2022	31.31	30.09	1.04	28.28	25.58	1.11
21/07/2022	30.04	29.31	1.02	28.83	26.00	1.11
22/07/2022	31.47	29.70	1.06	28.23	25.37	1.11
23/07/2022	–	–	–	28.17	26.16	1.08

*Where 'T_{max}' and 'T_{min}' account for maximum and minimum temperature, respectively, '*S*' means South, '*N*' is North and 'Ext' is exterior

Table 10 Statistical analysis of wind speed–temperature data gathered and collected by all sensors, respectively

	WS versus T Sensor1	WS versus T Sensor2	WS versus T Sensor3
Covariance	– 0.78	– 0.64	0.00
Correlation coefficient	– 0.49	– 0.61	0.00

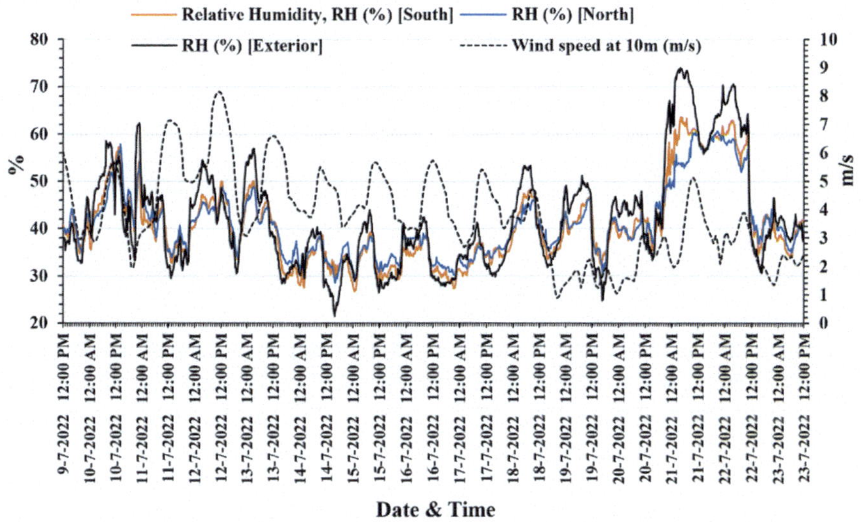

Fig. 15 Case study monitoring: indoor and outdoor ambient relative humidity and wind speed. *Source* sensors 1, 2, and 3, and NASA's 'POWER Project' Data Access Viewer, respectively [21]

Statistical data is also provided in Table 11 below:

Therefore, it is patent that both indoor relative humidities behave in accordance with outdoor data, with a higher correspondence of data recorded by Sensor 1_South, a room that is more exposed to the exterior.

Finally, in terms of wind activity, as seen in Fig. 15, its relationship with relative humidity data from all sources (sensors) becomes more intricate. Table 12 statistically analyses their connection.

While indoor relative humidity data follow a slightly different distribution in comparison with wind speed, outdoor data from Sensor ID 3_Exterior is clearly against its behaviour, showing an indirect relationship that may imply relative humidity drops as the wind increases.

Table 11 Statistical analysis of relative humidity data collected by all sensors

	RH Sensor 3 versus 1	RH Sensor 3 versus 2
Covariance	87.51	72.76
Correlation coefficient	0.94	0.91

Table 12 Statistical analysis of MERRA-2 wind speed and relative humidity data collected by all sensors

	WS versus RH Sensor 1	WS versus RH Sensor 2	WS versus RH Sensor 3
Covariance	– 0.62	– 0.79	– 2.40
Correlation coefficient	– 0.05	– 0.07	– 0.14

5 Conclusions

The adaptation of building occupants to unfavourable outdoor environmental conditions resides in two primary aspects: the passive strategies implemented in the building, comprising design features and subject to construction quality, and strategies of use, such as the control of window openings, sun protection solutions, among others inherent to their behaviour.

As far as the authors of this chapter know, this is the first time the hygrothermal performance of a building has been analysed under a series of unfavourable conditions affecting both indoor and outdoor ambient. In particular, considering the Mediterranean climate in summer, these conditions consisted of (i) keeping windows open during a high-temperature season and (ii) a nearby wildfire, which had a clear effect on the area's moisture index and could have impacted recorded relative humidity.

The case study building was monitored for two weeks in July 2022 in three key locations using temperature and relative humidity sensors (Sensor 1_South, 2_North, and Sensor ID 3_Exterior). The results indicated that the case study building's thermal inertia made it retain heat for 2–4 h from the logical temperature drop at night, with the most exposed room (south bedroom) reaching an average 4.40% reduction in temperature in relation to the mean maximum outdoor temperature; likewise, the indoor temperature rise took from 1 to 2 h to occur during sun activity from early morning. The non-sunlit north guest room showed a 10% higher average relative minimum daily indoor temperature (circa 28 °C) in comparison with the outside, thus revealing worse thermal behaviour (greater thermal discomfort) than the south bedroom at night. In addition, window closing had an impact on both indoor temperature and relative humidity inasmuch as their rises and drops happened at a slower pace, as expected. Similarly, their values increased as wind speed decreased, the latter following cycles of 12 h against the daily cycles (24 h) of temperature and humidity. For its part, building occupation in the south bedroom caused relative humidity to be slightly higher than that of the north guest room. However, the less exposed room (North) presents higher relative humidity despite the lack of occupation. In this sense, having installed Sensor 3 also on the north façade, it was not possible to evaluate the effect of watering the garden on the case study's indoor hygrothermal conditions.

All considered, future work would involve researching the role of indoor air velocity in temperature and relative humidity changes. Similarly, the impact of closing the window of the south bedroom during the day (the most affected room at a critical moment) on the building's hygrothermal behaviour is yet to be explored.

Acknowledgements This work has been supported by funding for a post-doctoral researcher contract from the VI Plan Propio de Investigación y Transferencia of Universidad de Sevilla (reference VIPPIT-2020-II.5), Spain.

This work has also been funded by the England European Regional Development Fund (ERDF) (reference 08R20S04177) as part of the European Structural and Investment Funds Growth Programme 2014-2020, the University of Nottingham, and Nottingham Trent University through the 'Live Experiential and Digital Diversification—Nottingham' (LEADD:NG) project.

The authors also wish to thank the Product Innovation Centre at the School of Architecture, Design and the Built Environment, Nottingham Trent University, for access to their sensors.

Finally, as stated in this chapter, third-party satellite data were obtained from the NASA Langley Research Center (LaRC) POWER Project, which is funded through the NASA Earth Science/ Applied Science Program.

References

1. Arsad FS, Hod R, Ahmad N et al (2023) Assessment of indoor thermal comfort temperature and related behavioural adaptations: a systematic review. Environ Sci Pollut Res 30:73137–73149. https://doi.org/10.1007/s11356-023-27089-9
2. Balali A, Yunusa-Kaltungo A, Edwards R (2023) A systematic review of passive energy consumption optimisation strategy selection for buildings through multiple criteria decision-making techniques. Renew Sustain Energy Rev 171:113013. https://doi.org/10.1016/j.rser.2022.113013
3. Chvatal KMS, Corvacho H (2009) The impact of increasing the building envelope insulation upon the risk of overheating in summer and an increased energy consumption. J Build Perform Simul 2:267–282. https://doi.org/10.1080/19401490903095865
4. Kim A, Wang S, Kim J-E, Reed D (2019) Indoor/outdoor environmental parameters and window-opening behavior: a structural equation modeling analysis. Buildings 9:94. https://doi.org/10.3390/buildings9040094
5. Mochida A, Yoshino H, Takeda T et al (2005) Methods for controlling airflow in and around a building under cross-ventilation to improve indoor thermal comfort. J Wind Eng Ind Aerodyn 93:437–449. https://doi.org/10.1016/j.jweia.2005.02.003
6. Daghigh R (2015) Assessing the thermal comfort and ventilation in Malaysia and the surrounding regions. Renew Sustain Energy Rev 48:681–691. https://doi.org/10.1016/j.rser.2015.04.017
7. Tamaskani Esfahankalateh A, Farrokhzad M, Saberi O, Ghaffarianhoseini A (2021) Achieving wind comfort through window design in residential buildings in cold climates, a case study in Tabriz city. Int. J. Low-Carbon Technol. 16:502–517. https://doi.org/10.1093/ijlct/ctaa082
8. Li N, Li J, Fan R, Jia H (2015) Probability of occupant operation of windows during transition seasons in office buildings. Renew Energy 73:84–91. https://doi.org/10.1016/j.renene.2014.05.065
9. Lai D, Jia S, Qi Y, Liu J (2018) Window-opening behavior in Chinese residential buildings across different climate zones. Build Environ 142:234–243. https://doi.org/10.1016/j.buildenv.2018.06.030
10. Zaki WR, Nawawi AH, Ahmad SS (2012) Environmental prospective of passive architecture design strategies in terrace houses. Proced Soc Behav Sci 1(42):300–310
11. Borghero L, Clèries E, Péan T et al (2023) Comparing cooling strategies to assess thermal comfort resilience of residential buildings in Barcelona for present and future heatwaves. Build Environ 231:110043. https://doi.org/10.1016/j.buildenv.2023.110043
12. Instituto de Estadística y Cartografía de Andalucía (2023) Alhaurín de la Torre. In: Andalucía pueblo a pueblo—Fichas Municipales

13. Google Inc (2023) Google Earth Pro
14. Dirección General de Protección Civil y Emergencias (2023) Informe de incendios forestales 2022
15. Dirección General de Política Forestal y Biodiversidad (2023) Resolución de la dirección general de política forestal y biodiversidad por la que se aprueba el plan integrado de caza de los terrenos afectados por el incendio de Mijas en el año 2022. Consejería de Sostenibilidad, Medio Ambiente y Economía Azul (Junta de Andalucía)
16. Acaire (2022) Seguimiento del incendio de la Sierra de Mijas de julio de 2022. In: Incendios en la Costa del Sol: Sierra de Mijas
17. Ministerio de Vivienda (2006) Real decreto 314/2006, de 17 de marzo, por el que se aprueba el Código Técnico de la Edificación. Boletín Oficial del Estado, Spain
18. European Space Agency (ESA) (2017) T-Systems Copernicus Browser. In: Copernicus Data Space Ecosystem
19. Antón D, Amaro-Mellado J-L (2021) Engineering graphics for thermal assessment: 3D thermal data visualisation based on infrared thermography, GIS and 3D point cloud processing software. Symmetry 13:335. https://doi.org/10.3390/sym13020335
20. Antón D, Carretero-Ayuso MJ, Moyano-Campos J, Nieto-Julián JE (2022) Laser scanning intensity fingerprint: 3D Visualisation and analysis of building surface deficiencies. In: Bienvenido-Huertas D, Moyano-Campos J (eds) New technologies in building and construction. Springer Nature Singapore, Singapore, pp 207–223
21. NASA (1981) Power data access viewer. In: Prediction of worldwide energy resource
22. NASA (2020) Power data methodology. In: NASA POWER

Thermal Vulnerability Analysis in Low-Income Housing with Natural Ventilation in the Galápagos Islands: Measured and Simulated Data

Santiago Navarro, Catalina Vallejo Coral, Marco Orozco Salcedo, Hugo Zúñiga Puebla, and Juana Perlaza

Abstract The Galapagos Islands are the world's second-largest marine reserve, declared a World Heritage Site in 1978 by UNESCO, and are located in a hot and humid climate zone. Their fragile ecosystem is susceptible to the adverse effects of global warming. The increase in ambient temperature can lead to a rise in the energy requirement to achieve thermal comfort in the residential sector, mainly made up of naturally ventilated dwellings. Several studies have been developed for Ecuador regarding the impact of passive strategies in naturally ventilated dwellings on the continent, but no studies have been conducted in the Archipelago. In this sense, the present study aims to evaluate the vulnerability of a naturally ventilated house in the Galapagos, using adaptive models based on temperature and thermal stress index based on temperature and humidity. The results showed adequate levels of thermal comfort, according to UNE-EN 16,798–1:2020. However, the heat index revealed that for approximately 40% of the time the dwelling was occupied, the occupants were highly vulnerable (32 °C < HI < 52°), increasing the likelihood of heat cramps, heat exhaustion, and heat stress.

S. Navarro · C. Vallejo Coral (✉) · M. Orozco Salcedo
Instituto de Investigación Geológico y Energético, Quito 170518, Ecuador
e-mail: catalina.vallejo@geoenergia.gob.ec

C. Vallejo Coral · H. Zúñiga Puebla
Departamento de Ingeniería Mecánica, Escuela Politécnica Nacional, Ladrón de Guevara E11 253, Quito, Ecuador

J. Perlaza
Department of European and Mediterranean Culture: Architecture, Environment and Cultural, University of Basilicata, Vía Lanera, 20, 75100 Matera, Italy

M. Torres González and C. Rubio Bellido (eds.), *Towards Low and Positive Energy Buildings*, Green Energy and Technology, https://doi.org/10.1007/978-3-031-70851-0_8

1 Introduction

A fundamental aspect of achieving balance between humans and their environment is ensuring physical well-being. Temperature and humidity are critical variables for human comfort, especially in dwellings with intermediate conditions where the perception of comfort is subjective. One individual may find the environment comfortable, while another may perceive it as too hot or too cold. This discrepancy arises because thermal sensation is influenced by subjective factors such as physiological and psychological characteristics. Nonetheless, maintaining the body's overall thermal balance is essential and depends on thermo-hygrometric conditions, clothing, and physical activity. Globally, the increase in extreme heat has led to higher rates of human mortality and morbidity. These phenomena have been driven by the expansion of urban areas, rising temperature extremes, and demographic changes [1]. Furthermore, it is projected that average ambient temperatures will rise by 1.5 °C over the coming two decades [2] posing a challenge to the thermal vulnerability of dwellings. In this scenario, Tavakoli et al. [2] emphasize the necessity for research into how occupants respond to thermal stress, particularly in the context of climate change. Thus, this study focuses on a literature review of the behavior of naturally ventilated dwellings as a potential solution for addressing forthcoming climate challenges. Additionally, it explores the critical need to comprehend physiological responses to heat in light of increasing global temperatures and their implications for human health, including psychological aspects. The study asserts that while naturally ventilated dwellings offer a viable option, integrating passive thermal comfort strategies is essential to mitigate the expected rise in heat wave temperatures, which are anticipated to lead to elevated mortality rates.

In a study by May et al. [3], in the tropical zone, particularly in the town of Myanmar, they evaluate the thermal behavior of a house, using a monitoring system and energy simulation considering future climate change scenarios. It mentions an increase in the number of hours with temperatures above 30 °C, compared to the same time interval during the previous year, and a higher percentage of time with a heat index. In these conditions, it is crucial to consider the risk of overheating to which the inhabitants are exposed. Gamero-Salinas et al. [4] analyzed the risk of overheating in different social dwellings during the hot season in a tropical climate in Tegucigalpa, Honduras. This study analyzes whether the risk of overheating is modified according to the building typology, the methodology of the CIBSE TM52:2013 standards [5], and the thermal comfort approach of the UNE-EN 16,798–1:2020 [6] and ANSI/ASHRAE 55:2017 [7] standards. It concludes that adaptive models in naturally ventilated dwellings are not sufficient to reduce vulnerability and that it is necessary to consider other options, such as passive strategies, to reduce the risk of overheating.

In Ecuador, a country located in the tropical zone, several studies have been carried out to analyze thermal comfort on the continent. Delgado-Gutierrez et al. [8] analyzed adaptive thermal comfort using the ASHRAE 22–2020 standard as a reference. Through simulation, future scenarios are evaluated considering the effects of climate change in 2050 and 2100. It concludes that thanks to the low variability of climatic conditions, adaptive comfort models can be applied using natural ventilation, particularly in low-income housing. In the second part of the study, Delgado-Gutierrez et al. [9] analyze the thermal comfort in social housing in several areas of Ecuador through simulations considering future scenarios to 2050, finding several comfort ranges depending on the area's climate. Vallejo et al. [10] evaluate the thermal performance of a social dwelling in Ecuador's hot-humid zone and question whether the current local NEC-HS-EE [11] standard is sufficient to guarantee adequate comfort conditions in residences in a hot-humid zone. In this case, authors conclude that although the house complies with the mentioned standard, it is necessary to define comfort criteria according to the local environmental conditions. In addition, the ambient temperature was evaluated using adaptive models according to ASHRAE 55 and EN 16,798–1, finding compliance with the latter; however, it was found that there is an increased risk of overheating of 13.9% according to ASHRAE 55.

Despite studies conducted in the Ecuadorian mainland's hot and humid tropical climate zones, there is a lack of information about the thermal vulnerability of dwellings in the Galapagos Archipelago. According to consultancy developed by the Bariloche Fundation [12], 60.8% of homes in the Galapagos Islands belong to the house/villa residential group. These buildings are naturally ventilated and are constructed with low thermal performance materials, roofs with high thermal transmittance, facades with low thermal resistance and high thermal inertia, and windows without adequate shading elements. In addition, many operating hours were observed where temperatures fell outside comfort conditions [2]. Integrating the urban metabolism analysis reveals a high dependence on external energy resources, exacerbating thermal vulnerability. Sustainable design practices, such as improving building insulation, incorporating reflective roofs, and promoting natural ventilation, are crucial to increasing resilience. The use of Building Information Modeling (BIM) in the study by Colmenero et al. [13] enabled the creation of accurate simulations of the thermal performance of low-income housing. BIM facilitates evaluating and optimizing various sustainable design strategies, such as improving building insulation, incorporating reflective roofs, and promoting natural ventilation. These strategies enhanced thermal comfort, reduced energy consumption, and increased the sustainability of low-income housing in hot and humid climates such as the Galapagos Islands.

In this context, it is necessary to thoroughly analyze the applicability of regulations that do not consider the natural environmental conditions of dwellings located in the tropical zone since there are variables that have not been considered and are determinants for the thermal comfort of human beings. Vellei et al. [14] mention that current regulations only indicate a linear relationship between indoor and outdoor temperatures without considering other variables such as humidity or airspeed. They demonstrated that humidity must exist in an adaptive thermal comfort model and shows that the conventional model overestimates overheating by 30% concerning the new analysis considering humidity. In this scenario, the present study seeks to assess the vulnerability of a naturally ventilated dwelling in the Galapagos, using adaptive models based on temperature and thermal stress index based on temperature and humidity.

2 Methodology

2.1 Case Study

In the Galapagos Archipelago, 50% of the dwellings are one-story, single-family houses with metal or fiber cement roofs and without air conditioning. The main problems identified are the roof's high thermal transmittance, facades with high thermal inertia, and windows without shading [12].

The case study presents the same problem with the materials described in Table 1 and the architectural configuration presented in Fig. 1. Specifically located on Santa Cruz Island (0°44′00.5 "S 90°18′48.8 "W) (GMT-6.0 Hour), the house occupies a construction area of 109.62m2. It comprises one bedroom without a bathroom, three bedrooms each with private bathrooms, a work room, a social room, and a kitchen, as illustrated in Fig. 2. The work room features an exterior galvanized steel roof without a false ceiling, while bedroom 4 has a fiber cement roof with a false ceiling; other areas have galvanized steel roofs with false ceilings. Windows consist of mesh, facilitating continuous natural ventilation. Situated two meters below street level, the house includes a cyclopean wall at the front and is flanked by two adjacent residential buildings: a two-story structure to the east and a four-story building to the west at street level.

It is unoccupied from 10:00 am. to 5:00 pm. Monday through Friday and occupied 24 h a day on weekends. The dwelling has five occupants who wear summer clothing, representing a thermal resistance of 0.5clo and a metabolic rate of 180W/person [7].

2.2 Meteorological Data

Santa Cruz Island has a maximum altitude of 864 m.a.s.l. and is the second-largest island in the Galapagos Archipelago, with a population of 15,700 inhabitants [19]. According to the Köppen-Geiger climate classification, Santa Cruz belongs to the Aw group (tropical savannah climate) [20]. It experiences two distinct seasons: a warm, humid season with average temperatures ranging from 23 to 27 °C, and a cold, dry season where temperatures typically range between 20 and 22 °C, as depicted in Fig. 3. The prevailing winds come from the southeast, as shown in Fig. 4, intensifying in the cold season and weakening in the warm season [21], and the average monthly humidity has fluctuated between 75 and 85% in the last five years [22].

Table 1 Material properties of the building envelope of the studio house

Element	Material	Thermal conductivity [W/m °C]	Density [kg/m^3]	Specific Heat [J/kg °C]	Solar Absorptance	Emissivity	Thickness [mm]
Roof	Aluminum [15]	160	2,800	880	0.5	0.3	0.3
	Fiber-cement [16, 17]	0.93	2,000	1,250	0.7	0.9	5
	Gypsum ceiling [16, 17]	0.25	900	1,000	0.5	0.9	15
Walls	Concrete blocks [17, 18]	0.81	977	837	0.6	0.9	140
	Cement plaster [17]	1.35	1,800	1,000	0.6	0.9	10
	Stucco [17]	0.76	1,000	837	0.7	0.9	5
Windows	Mesh	–	–	–	–	–	–
Door	Wood (Laurel) [17]	0.13	650	1,700	0.6	0.9	40
Floor	Concrete [17]	1.35	1,800	1,000	0.6	0.9	150
	Ceramic tiles [17]	0.84	1,900	921	0.65	0.9	10

Fig. 1 Santa Cruz Island studio housing

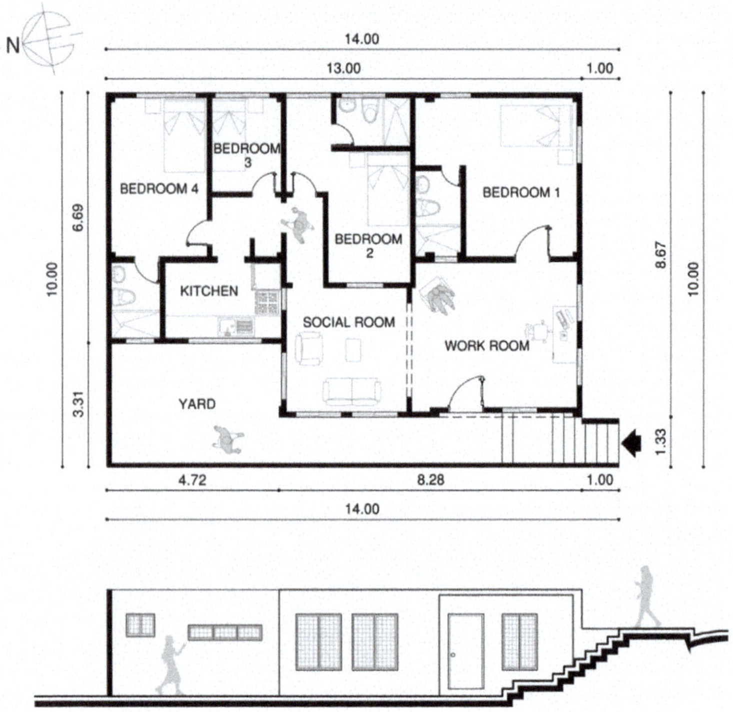

Fig. 2 Plan and elevation of the case study: House in Santa Cruz (Galapagos, Ecuador)

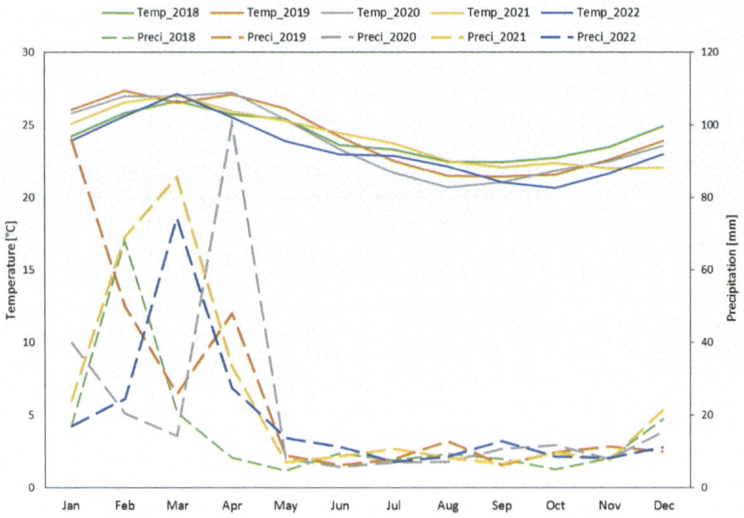

Fig. 3 Historical ambient temperature and precipitations in Santa Cruz from 2018 to 2022 [22]

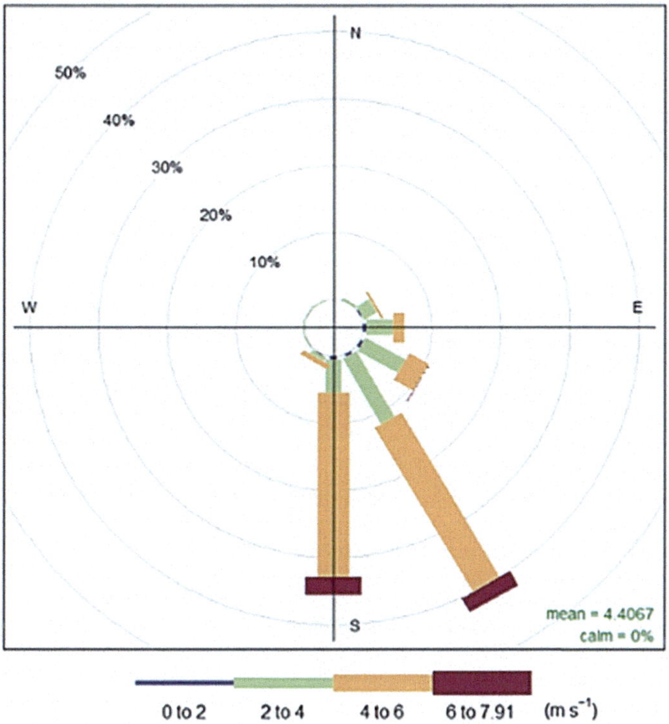

Fig. 4 Historical monthly wind speed and direction (2018–2022) in Santa Cruz [22]

2.3 *Energy Simulation*

The software used for the energy simulation of the study house was Design builder version 4.7.0.027. The software has a built-in Energy-Plus engine and functionalities like 3D modeling, evaluation, and optimization. The house model shown in Fig. 5 was designed according to the construction plans. To obtain the closest results to reality, it is necessary to calibrate the model [23] using precise data on the internal air temperature of the house.

2.3.1 Monitoring

As shown in Fig. 6, temperature and relative humidity measurements were recorded in the social room and the workroom from 27/04/2023 to 20/07/2023 (84 days), at five minutes intervals. An ONSET thermohygrometer, model HOBBO UX100-003, was installed in each space. The sensors were placed at a height of 1.7 m, with an accuracy of ± 0.21 °C for temperature and ± 3.5% for relative humidity.

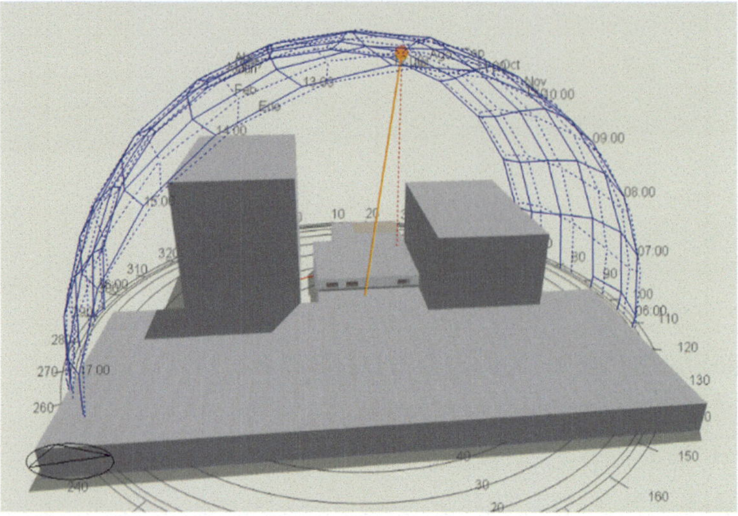

Fig. 5 Model of the study house in the DesignBuilder software

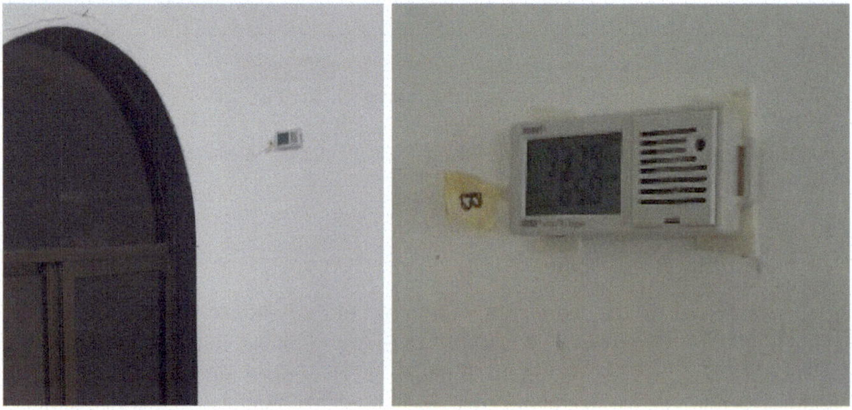

Fig. 6 Temperature and relative humidity monitoring in a studio flat

Meteorological data were gathered every 10 min from a weather station operated by the Charles Darwin Foundation in Puerto Ayora (0°44′37.6"S 90°18′07.2"W) [24]. The parameters measured were air temperature, wind speed and direction, precipitation, relative humidity, solar radiation, and soil temperature [24].

2.3.2 Calibration

For the calibration of the model, the climate file for the monitoring period was obtained in Energy Plus Weather (EPW) format from the actual data collected by the on-site weather station. The model's validity is determined using the methodology described in ASHRAE Guideline 14–2014 [25] and Mutasim et al. [26] employed various statistical parameters including the coefficient of variation of the root mean square error (CV [RMSE]) calculated using Eq. (1) and the normalized mean trend error (NMBE) calculated using Eq. (2). To calibrate the model, and thus represent the thermal behavior of the dwelling, the values of the CV [RMSE] and NMBE should be a maximum of 30% and 10%, respectively. To avoid discrepancy between the measured and simulated values of the internal temperature of the dwelling, when the CV [RMSE] and NMBE comply with the permissible ranges, the maximum absolute difference (MAD) is calculated with Eq. (3). The mean absolute deviation (MAD) criterion identifies the maximum discrepancy between measured and simulated values over the analysis period. Meanwhile, the percentage error criterion of ± 1 °C measures the proportion of hours where the temperature difference between actual and simulated values exceeds one degree. Given the significance of a 1 °C difference in determining overheating hours, this calculation is crucial for minimizing discrepancies in assessment [26]. There is no standard specifying acceptable ranges for MAD indicators and the ± 1 °C percentage error criterion. However, in his study, Mutasim [26] achieved a MAD of less than 2 °C and a ± 1 °C percentage error criterion of less than 10% for a satisfactorily calibrated dwelling.

$$\mathrm{CV[RMSE]}(\%) = \frac{100}{\overline{T_m}} \cdot \left[\frac{1}{n-1} \cdot \sum_{j=1}^{n} \left(T_{sj} - T_{mj}\right)^2 \right]^{0.5} \tag{1}$$

$$\mathrm{NMBE}(\%) = \frac{100}{\overline{T_m}} \cdot \frac{\sum_{j=1}^{n} T_{sj} - T_{mj}}{n-1} \tag{2}$$

$$\mathrm{MAD}(^\circ\mathrm{C}) = \max\left(\mathrm{abs}\left(T_{sj} - T_{mj}\right)\right),\ j = 1 \ldots n \tag{3}$$

where J is the specific time during the calibration period, T_s is the simulated temperature, T_m is the measured temperature, n is the number of observations, and $\overline{T_m}$ is the average temperature measured over n observations.

2.4 Thermal Performance Assessment

To understand the dwelling's thermal performance, the adaptive approach was considered. This approach defines the maximum operating temperature acceptance limits and assesses the risk of overheating to measure its severity and occurrence [5].

In addition, the heat index temperature was calculated to identify the influence on thermal comfort.

2.4.1 Adaptive Model of Thermal Comfort

Thermal comfort in naturally ventilated dwellings is assessed using the adaptive models set out in ASHRAE 55 and EN 16,798–1. The models establish the comfort limit for the internal operating temperature. Equation (4) corresponds to the upper limit with 80% acceptability according to ASHRAE 55 and is used for typical applications. Equation (5) corresponds to the upper limit of Category II according to EN 16,798–1. The exponentially weighted running mean of the outdoor air temperature (*T*_rm) is calculated according to Eq. (6).

$$T_{\max}(80\%\ \text{acceptability}) = 0.31 \cdot T_{rm}(^{\circ}\text{C}) + 21.3(^{\circ}\text{C}) \tag{4}$$

$$T_{\max}(\text{Category}II) = 0.33 \cdot T_{rm}(^{\circ}C) + 21.8(^{\circ}C) \tag{5}$$

$$T_{rm} = (1 - \alpha) \cdot T_{ed(n-1)} + \alpha \cdot T_{rm-1} \tag{6}$$

where T_{rm} is the exponentially weighted moving average of the outdoor air temperature, T_{ed} is the average daily outdoor temperature, (n-1) refers to the previous day's parameter, and α is the constant of the rate at which the operating temperature changes concerning the outside temperature for humid tropical climates is 0.9 [7].

2.4.2 Risk of Overheating

The limits defined in the previous section are valid for identifying the occurrence of overheating and average or cumulative values. However, they do not allow for assessing its severity [2, 5]. The Technical Memorandum standard TM52: 2013, from the Chartered Institution of Building Services Engineers (CIBSE), outlines a methodology for assessing overheating in dwellings by using three criteria in terms of the difference between the indoor operating temperature and the maximum acceptable temperature ΔT as presented in Eq. (7).

$$\Delta T = T_{\text{op}} - T_{\max} \tag{7}$$

2.5 Criterion 1 (C1): Hours of Leave (He)

The percentage of occupancy hours where ΔT is equal to or greater than one degree (°C) and must not exceed 3% is expressed in Eq. (8).

$$\text{He} \leq 3\% \tag{8}$$

2.6 Criterion 2 (C2): Daily Weighted Excess Hours (We)

The time (hour or fraction of an hour) during which the operating temperature exceeds the maximum temperature during the hours of occupancy is weighted by a factor that is a function of how many degrees are above the range. Equation (9) is used to calculate it. We must be equal to or less than 6 for any day of the period, as expressed in Eq. (10).

$$\text{We} = \sum h_e x \, \text{WF} \tag{9}$$

The weighted factor is $\text{WF} = 0$ if $\Delta T \leq 0$; otherwise, $\text{WF} = \Delta T$, and h_eyes the number of hours where $\text{WF} = y$.

$$\text{We} \leq 6 \tag{10}$$

2.7 Criterion 3 (C3): Upper Limit Temperature (Tupp)

This is the absolute maximum temperature for the indoor operating temperature. ΔT shall not exceed 4 °C, as expressed in Eq. (11).

$$\Delta T \leq 4(^{\circ}\text{C}) \tag{11}$$

A dwelling is classified as overheated when it fails to meet two out of these three criteria.

2.7.1 Heat Index Temperature

Humans require specific thermal conditions in the environment to maintain thermal comfort. However, if the surrounding environment has temperatures above 32 °C and high relative humidity, the body loses its ability to cool itself and consequently experiences negative health effects from high temperatures [3]. The heat index temperature (HI) is a widely used metric to assess health risk (heat stress) by considering relative humidity and air temperature [2]. Equation (12) is a multiple regression developed by Lans P. Rothfuss based on hourly data. Suppose the relative humidity is greater than 85% and the temperature is between 26.67 (80 °F) and 30.56 °C (87 °F). In that case, adding the factor calculated with Eq. (13) is necessary, corresponding to a Fahrenheit temperature scale adjustment. If HI is less than 26.67 °C (80 °F), Eq. (12) is not appropriate, so Eq. (14) is used [3]. Equations (12), (13), and (14) determine the HI in Fahrenheit, and to obtain the results in °C, the corresponding unit conversion is performed.

$$\begin{aligned} HI(^{\circ}\mathrm{F}) = &- 42.379 + 2.04901523 \cdot T + 10.14333127 \cdot R - 0.22475541 \cdot T \cdot R \\ &- 0.00683783 \cdot T^2 - 0.05481717 \cdot R^2 + 0.00122874 \cdot T^2 \cdot R \\ &+ 0.00085282 \cdot T \cdot R^2 - 0.00000199 \cdot T^2 \cdot R^2 \end{aligned} \tag{12}$$

where T is the dry bulb temperature (°F) and R is the relative humidity in percent.

$$\mathrm{HI}(^{\circ}\mathrm{F}) = \left(\frac{R - 85}{10}\right) \cdot \left(\frac{87 - T}{5}\right) \tag{13}$$

$$\mathrm{HI}(^{\circ}\mathrm{F}) = 0.5 \cdot \{T + 61 + [(T - 68) \cdot 1.2] + (R \cdot 0.094)\} \tag{14}$$

Heat indices are categorized into different risk scales, as detailed in Table 2.

Table 2 Heat hazard classification and impact on the body [27]

Heat index [°C]	Category	Effect on the body
27–32	Caution	Fatigue is possible with prolonged exposure and activity—heat cramps
32–39	Extreme Caution	Heat cramps and heat exhaustion are possible—heatstroke
39–51	Danger	Heat cramps and heat exhaustion are likely; heatstroke is probable with continued activity
Above 52	Extreme Danger	Heatstroke is imminent

3 Results

3.1 Calibration and Energy Balance

To perform the calibration of the energy model, air temperature and relative humidity data were collected for 84 days in the social room of the dwelling. The model complied satisfactorily with the statistical indicators described in Calibration section, and the results are presented in Table 3. The CV [RMSE] and NMBE were 2% and 1.31%, respectively. The model has minimal bias and low variability in its predictions. The ± 1 °C error is 2.11%, and the most significant difference between actual and simulated temperature occurs most frequently around midday when radiation is highest. Based on the results given in Table 3, the model complies with the calibration parameters and adequately represents the actual thermal behavior of the house.

Figure 7 presents the mean daily heat balance for the year 2023. During the hours of solar radiation (6:00am–18:00 pm), the roof has the highest thermal gain, presenting its maximum value at midday due to the high solar radiation. The thermal loads show similar behavior through the windows. For this reason, the internal temperature increases around midday, as shown by studies carried out by [9, 28], and [29] in dwellings in similar climatic zones. The house has no glazing in the windows, which is why the model is highly sensitive to the thermal properties of the roof, which are defined according to the literature. The heat gains from the walls increase in the afternoon, and during the morning, they have heat losses. This may be due to their high thermal inertia [12], absorbing heat during the morning and releasing it in the afternoon and evening hours to the interior of the dwelling. There are heat gains due to occupancy and lighting during the hours when the dwelling is occupied. Finally, the floor is the only element of the envelope through which there are no heat gains, and the most significant heat gains occur from 7:00 am to 18:00 pm.

Figure 8 shows the indoor temperature measured and simulated during the 8 days of the monitoring period. During this period, on April 30, at 1 pm, the MAD is 1.86 °C, representing an error of 6.33%.

Table 3 Model calibration results

Parameter	Requirement	Simulated
CV [RMSE]	< 30%	2%
NMBE	< 10%	1.31%
MAD	< 2°C	1.86°C
± 1°C error	< 10%	2.11%

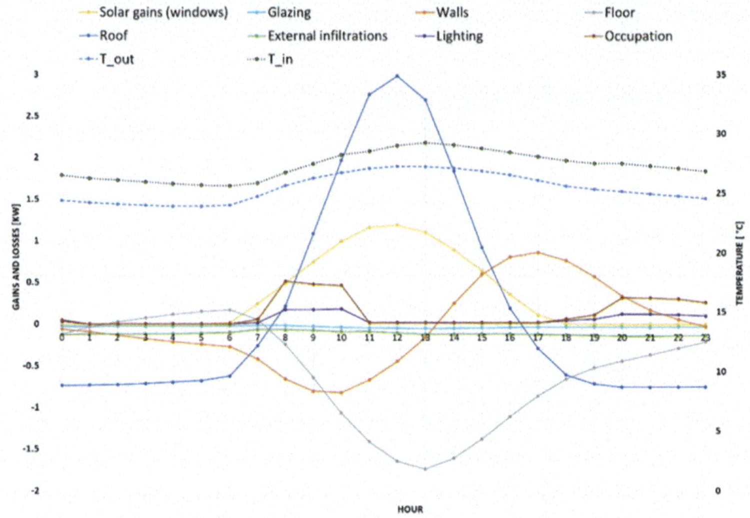

Fig. 7 Heat balance of an average day

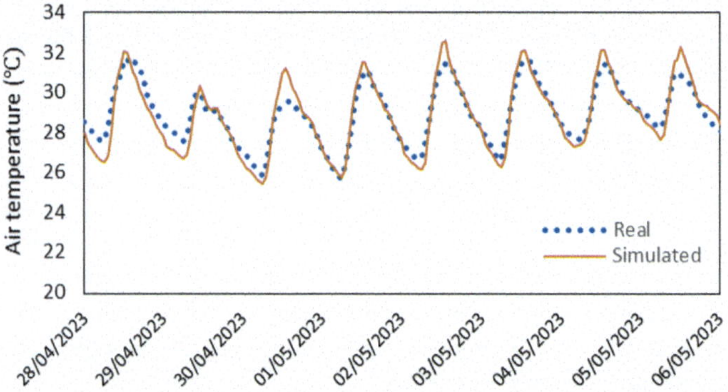

Fig. 8 Comparison of simulated and monitored internal temperature (28/04/2023–05/05/2023)

3.2 *Thermal Performance Assessment*

The thermal performance was evaluated using data obtained through energy simulation in DesignBuilder for 2023. For this purpose, the climate file (EPW) was obtained from the meteorological data for the same period recorded by the Charles Darwin Foundation station [21].

3.2.1 Adaptive Thermal Comfort Model

The comfort limits of EN 16,798–1—Category II represent an adequate level of comfort for the majority of occupants for most of the time (95% of occupancy hours) [6]. On the other hand, ASHRAE 55 sets limits for at least 80% of occupants to be comfortable during all hours of occupancy and ensures a higher level of comfort [7]. This is the reason why the maximum temperature defined with Eq. (4) is always lower than that obtained with Eq. (5), as evidenced in Tavakoli et al. [2]. During 2023, the dwelling was occupied for 6,939 h, according to the schedule described in the case study section. Figure 9 shows the percentage of hours of discomfort in which the operating temperature of each space in the dwelling exceeds the comfort limits set in ASHRAE 55 and EN 16,798–1 described in previous section. In bedroom 4, 16.2% of the occupied hours exceed the limits set in ASHRAE 55 and 9% EN 16,798–1, the most critical space. The work and social rooms have around 8% discomfort hours according to ASHRAE 55 and up to 4.4% according to EN 16,798–1. All additional spaces have up to 7.5% discomfort hours, according to ASHRAE 55.

According to the thermal balance of all spaces, bedroom 4 and the workroom have the highest thermal load through the ceiling (2.5 kWh) from 6:00 am to 18:00 pm. The working room has a metal ceiling without a false ceiling, while the ceiling of bedroom 4 is made of fiber cement together with a false ceiling, which delays the occurrence of the maximum temperature in bedroom 4 by 2 h. On the other hand, the operating temperature of bedroom 4 and the social room has the same behavior from 21:00 pm to 11:00 am because the highest thermal load is through the ceiling, which has a false ceiling. However, the north-facing wall of bedroom 4 is the façade with the most prolonged exposure to solar radiation. This causes thermal gains, which, together with the thermal inertia of the wall, leads to an increase in the operating temperature from 12:00 pm to 20:00 pm. Finally, the social room and the workroom have the highest solar load through windows (3.3 kWh and 1.94 kWh, respectively) because the west-facing wall is the façade with the highest incidence of solar radiation during the day. This means that these spaces have a higher percentage of hours of

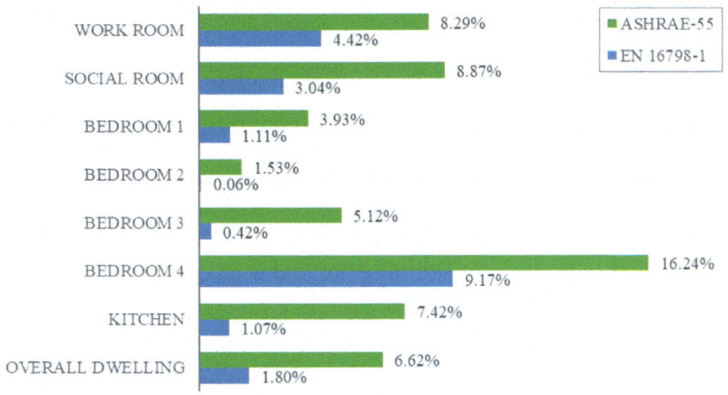

Fig. 9 Percentage of discomfort hours ASHRAE 55 and EN 16,798–1 during 2023

discomfort than the other spaces in the house. The energy simulation results indicate that the roof is the primary source of thermal gains due to prolonged sun exposure and low thermal inertia. This finding aligns with previous studies conducted in similar climatic zones, such as Thailand [30], the Dominican Republic [31], Indonesia [32], and the Ecuadorian mainland [33]. Side facades, which adjoin shaded buildings, and the floor, which experiences permanent heat loss, also significantly influence the indoor environment. The evaluation shows that, according to ASHRAE 55, more than 20% of the dwelling's occupants feel thermal discomfort in all spaces because the operating temperature is outside the maximum acceptable limits; based on EN 16,798–1, only bedroom 4 does not present acceptable thermal conditions. This is evidence of overheating in the dwelling, but its severity and associated health risks are unknown.

3.2.2 Overheating Risk

The results presented in Table 4 identify the severity of overheating. The dwelling has overheated in bedroom 4 because, during the year 2023, there are 34 days in which the daily weighted exceedance exceeds 6 h (i.e., it exceeds the acceptable limit of daily overheating). In addition, it does not meet criterion 3 because it exceeds the threshold above which occupants cannot adapt to achieve thermal comfort and will consequently feel too hot. The highest temperatures occur during the afternoon, as shown in Fig. 10 when the operating temperature exceeds the maximum allowable limit. Bedroom 4 cannot provide habitable conditions for heat waves or future extreme weather scenarios [5]. The work and social rooms have high thermal gains through the roof but lose heat through the walls during the morning due to the shade generated by the adjoining dwelling. In the social room, this causes the operating temperature to be lower in the morning hours, as shown in Fig. 10. For 11 days, the operating temperature exceeds the acceptable daily limit for overheating (criterion 2). The additional spaces do not risk overheating as they meet at least two assessment criteria and are considered habitable spaces, according to CIBSE TM52:2013 [5].

3.2.3 Heat Index Temperature

The results obtained in the previous sections show that occupants can adapt to the operating temperature and feel comfortable in most living spaces except for bedroom 4. The adaptive models used, as well as the assessment of the risk of overheating, focus on outdoor temperature and do not consider relative humidity and adaptive capacity. However, the moisture content of the air affects the human body's ability to regulate internal temperature through perspiration, negatively affecting health [3]. Kong et al. [34] identified that spaces with high humidity (relative humidity greater than 70%) elicit more robust thermal responses in occupants, such as increased thermal sensitivity, significant variations in wind chill, and skin temperature. These results

Table 4 Overheating risk assessment for the study dwelling scenarios in each bedroom

Dwelling overall summary	Criterion 1		Criterion 2		Criterion 3	
	Hours of Exceedance		Daily Weighted Exceedance		Upper Limit Temperature	
	He Max (3%)	Pass/Fail	We > 6	Pass/Fail	ΔT > 4 (°C)	Pass/Fail
Work room	2.22%	Pass	11	Fail	0	Pass
Social room	1.59%	Pass	3	Fail	0	Pass
Bedroom 1	0.12%	Pass	0	Pass	0	Pass
Bedroom 2	0.00%	Pass	0	Pass	0	Pass
Bedroom 3	0.09%	Pass	0	Pass	0	Pass
Bedroom 4	5.91%	Fail	34	Fail	15	Fail
Kitchen	0.16%	Pass	0	Pass	0	Pass

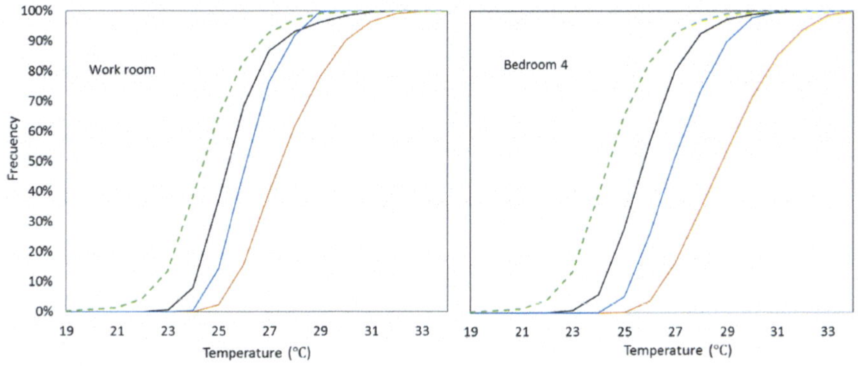

Fig. 10 Cumulative frequency of bedroom 4 and workroom temperature during the year 2023

show the importance of humidity in evaluating the vulnerability of the occupants in their spaces, especially in hot and humid places like Puerto Ayora.

In the studied dwelling, the relative humidity of the internal air is less than 70% only 10% of the time (of the year 2023) in bedroom 4 and the Workroom, as shown in Fig. 11. Furthermore, inside the dwelling, the relative humidity is lower than outside 65% of the time in areas such as bedroom 3 and kitchen and in the additional regions. This is because the indoor environment is always warmer than the outdoor environment (see Fig. 10), which increases the capacity of the air to hold more air vapor, causing the relative humidity to decrease [35].

The high relative humidity in the dwelling required a thermal vulnerability analysis combining temperature and relative air humidity to more accurately assess the occupants' actual thermal sensation. To identify whether occupants were exposed to conditions affecting their health during their stay in the dwelling in 2023, the HI was calculated using Eqs. (12), (13), and (14) considering the air temperature (dry bulb) and each space's simulated internal relative humidity. The HI results are presented in

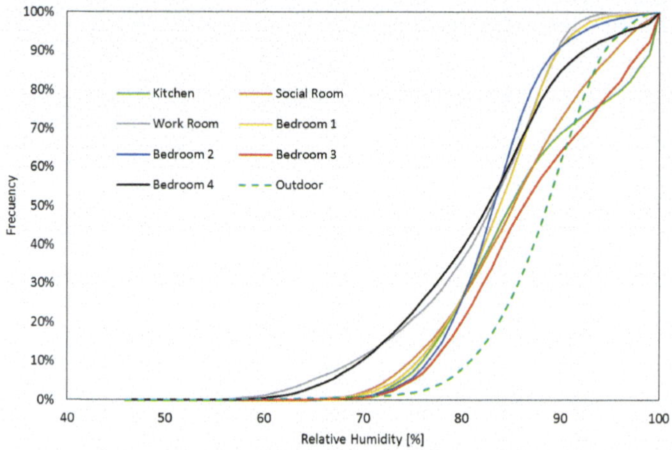

Fig. 11 Cumulative frequency of indoor relative humidity during the year 2023

Figs. 12 and 13. Throughout the day, variations of HI are observed from 22 °C in the workroom at 5:00 am and up to 54 °C in bedroom 4 at 10:00 am. During the hours when the occupants are not in the dwelling (10:00 am–17:00 pm), the HI reaches the highest values in all rooms, due to the increase of the internal temperature due to solar radiation and the high humidity of the external environment. From 20:00 pm onward, the kitchen and nearby areas increase the HI due to increased water in the environment due to the activities carried out. In addition, it is observed that during weekends (all-day occupancy), bedroom 4 is an extreme precaution zone (HI > 52 °C) and that all spaces except bedroom 2 and bedroom 1 are pre-hazard areas (39 °C < HI ≤ 52 °C) from midday onward.

Considering the occupancy time of the dwelling during the year 2023, Fig. 14 shows the percentage of hours in which the HI reaches the values according to the

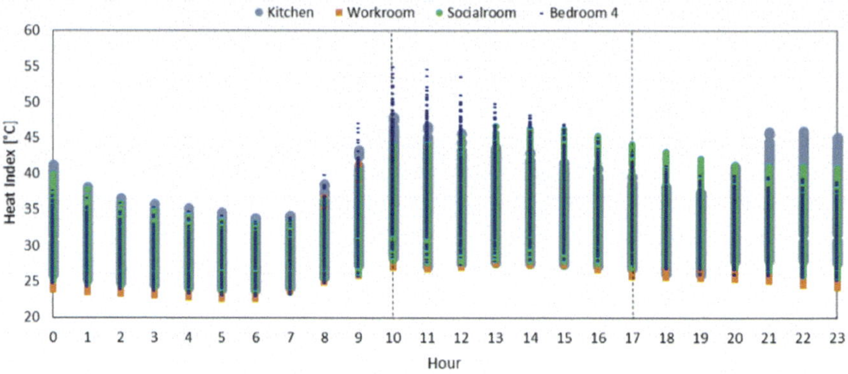

Fig. 12 Heat index of four housing zones during 2023

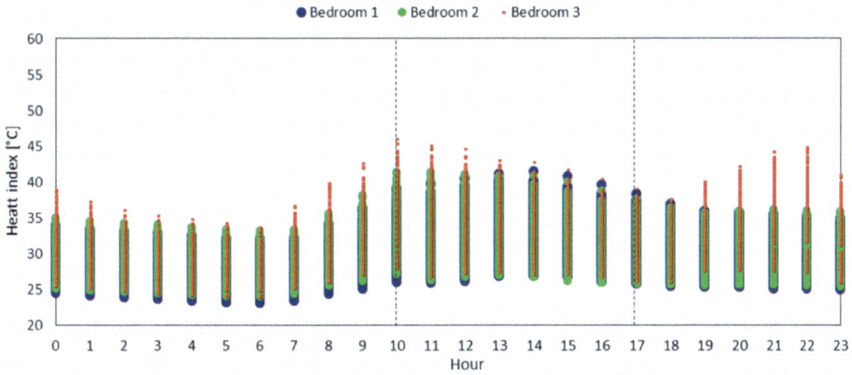

Fig. 13 Heat index of three housing zones during 2023

ranges described in Table 2. In the kitchen, room 3, the social room, and room 4, between 37% (2,567 h) and 31% (2,151 h) of the occupied time, the thermal environment represents a health hazard, with HI values between 32 and 39 °C; while between 4% on average (277 h), these areas present a hazardous thermal environment, which has adverse health effects such as cramps, exhaustion, and heat stroke.

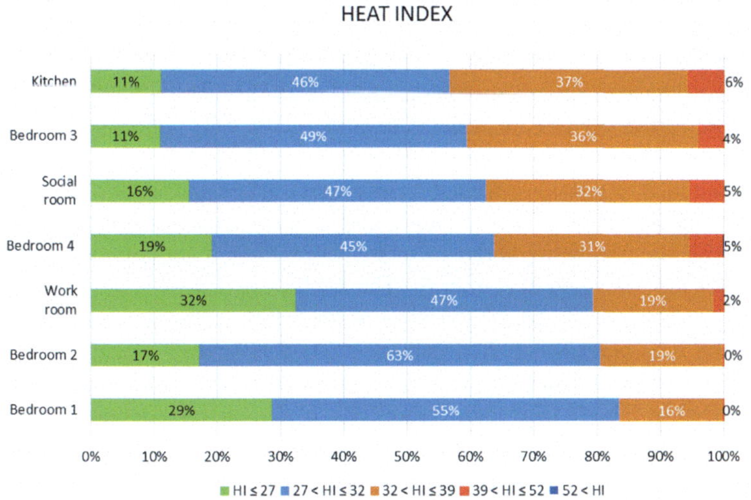

Fig. 14 Heat index temperature range of the simulated year 2023

4 Discussion

4.1 Analysis of Results

This study analyzes the vulnerability of a naturally ventilated house in Puerto Ayora in the Galapagos Archipelago. The house's internal and external conditions were monitored from April to July 2023 (84 days), energy calibration and simulation were performed, and different metrics such as thermal comfort, risk of overheating, and thermal stress were evaluated. Based on the calibrated model, the house simulation 2023 was carried out. The energy balance showed that the roof is the main element through which thermal gains exist due to its prolonged time to solar radiation and low thermal inertia. This result agrees with that identified in consultancy developed by the Bariloche Foundation [12]. The lateral facades of the house adjoin two buildings that generate shade along the entire solar trajectory (east–west). This causes the walls to be elements through which heat is lost, which is beneficial for lowering the operating temperature. Similarly, the floor is an element with permanent losses. In several studies, the same thermal behavior was shown for naturally ventilated dwellings in similar climatic zones such as Thailand [30], Dominican Republic [31], Indonesia [32], and mainland Ecuador [33]. The thermal behavior of the described dwelling directly influences the indoor environment. ASHRAE 55, EN 16,798–1-Category II, and CIBSE TM52 were used to assess the residence's thermal comfort and analyze overheating. The results showed that bedroom 4 is the most vulnerable area in the dwelling. The discomfort ranges vary between 9.17 and 16.24% because adaptive models consider the capacity of occupants to adapt to thermal environments that vary according to external conditions, generating a wide range of comfort depending on culture, people, and other factors [2]. However, this analysis does not consider relative humidity or provide information on the risk of heat stress for occupants. The heat index showed that the dwelling represents a risk of heat stress for the occupants. In all spaces, the lowest risk is related to possible fatigue if activities are performed on a prolonged basis and heat cramps. The kitchen, bedroom 3, social room, and bedroom 4 are the most critical spaces as at least 30% of the time occupants may suffer from heat cramps and heat exhaustion, and at least 5% of the hours may suffer from imminent heat stress if activities are carried out continuously, which is characterized by a body temperature above 40 °C. The thermal comfort assessment identified the vulnerability of the living areas and identified bedroom 4 as the space with the highest risk of overheating for the occupants.

However, the heat index showed the importance of relative humidity in assessing vulnerability in the dwelling due to the particularity of the climate zone. The adaptive models identify the risk of overheating in hot and humid climate zones but do not allow us to know the real impact of overheating in conjunction with relative humidity on the health of the occupants. The results' analysis reveals that the dwelling presents a significant risk of overheating, especially in bedroom 4 and the workroom. High thermal loads through the roof and windows contribute to many hours of thermal discomfort. These findings are consistent with previous studies in hot and humid tropical climates, where lack of adequate insulation and prolonged exposure to solar radiation increase the internal operating temperature of dwellings [30]. Furthermore, the results indicate that current standards, such as ASHRAE 55 and EN 16,798–1, may not be sufficient to ensure adequate comfort conditions in tropical areas with high relative humidity. Incorporating heat stress indices that consider relative humidity is crucial to accurately assess thermal comfort in these regions.

4.2 Implementing Sustainable Design Strategies

Integrating urban metabolism analysis reveals a high dependency on external energy resources, exacerbating thermal vulnerability. Sustainable design practices, such as improving building insulation, incorporating reflective roofing, and promoting natural ventilation, are crucial for enhancing resilience. The use of BIM in the study by Colmenero et al. [13] demonstrated its effectiveness in optimizing building performance and providing detailed insights into thermal behavior under various scenarios. These findings are particularly relevant for the Galápagos Islands, where our simulation results show significant overheating and high thermal loads through the roof. Adopting these strategies could significantly reduce thermal discomfort hours and the risk of heat stress, improving the habitability and sustainability of the housing. By incorporating better insulation, reflective roofs, and promoting natural ventilation, we can mitigate thermal stress and enhance the overall thermal comfort for the occupants. Future studies should consider implementing other passive and active heat mitigation strategies and more advanced monitoring systems to obtain more accurate data on thermal performance of dwellings. In addition, longitudinal studies are suggested to assess the long-term effects of the implemented strategies and their impact on occupants' health and well-being.

5 Conclusion

Naturally ventilated dwellings in hot and humid climate zones are inhabited by people from vulnerable populations who do not have the resources to purchase active cooling systems. In this sense, assessing the thermal vulnerability of this type of housing is crucial to know the comfort and health risks of the occupants, especially in the context of climate change that intensifies heat waves. The studied dwelling presents high vulnerability since, during the year 2023, the occupants presented a high risk of suffering from heat cramps, heat exhaustion, and heat stroke most of the time. However, the exclusive assessment of the risk of overheating and thermal comfort minimized this reality. In hot climate zones that are characterized by high relative humidity (greater than 70%), such as the Galapagos Archipelago, it is necessary to use metrics that take this variable into account, as body cooling through sweat evaporation is a natural mechanism for maintaining body temperature at healthy thresholds. The assessment of thermal vulnerability in naturally ventilated low-income housing in the Galápagos highlights the critical need for climate-responsive design. Better insulation, reflective roofing, and strategic ventilation can significantly mitigate thermal stress and improve occupant comfort. The findings underscore the importance of incorporating local climate conditions into building design to enhance sustainability and resilience.

References

1. Pérez De Ciriza PA (2018) Evaluación del Bienestar Térmico en Locales de Trabajo Cerrados Mediante los Índices Térmicos PMV y PPD
2. Tavakoli E, O'Donovan A, Kolokotroni M, O'Sullivan PD (2022) Evaluating the indoor thermal resilience of ventilative cooling in non-residential low energy buildings: a review. Build Environ 222:109376. https://doi.org/10.1016/j.buildenv.2022.109376
3. Zune M, Rodrigues L, Gillott M (2020) The vulnerability of homes to overheating in Myanmar today and in the future: a heat index analysis of measured and simulated data. Energy Build 223:110201. https://doi.org/10.1016/j.enbuild.2020.110201
4. Gamero-Salinas JC, Monge-Barrio A, Sánchez-Ostiz A (2020) Overheating risk assessment of different dwellings during the hottest season of a warm tropical climate. Build Environ. https://doi.org/10.1016/j.buildenv.2020.106664
5. Chartered Institution of Building Services Engineers (CIBSE) (2013) The limits of thermal comfort: avoiding overheating in European buildings. CIBSE Technical Memorandum 52 (TM52:2013). Great Britain
6. Asociación Española de Normalización (2020) UNE-EN 16798–1: Eficiencia energética de los edificios -Ventilación de los edificios—Parte 1:Parámetros del ambiente interior a considerar para el diseño y la evaluación de la eficiencia energética de edificios incluyendo la calidad del aire interior, co. Madrid
7. ASHRAE (2017) ANSI/ASHRAE 55: Thermal enviromental conditions for human occupancy
8. Delgado-Gutierrez E, Canivell J, Bienvenido-Huertas D, Hidalgo-Sánchez FM (2024) Adaptive comfort potential in different climate zones of Ecuador considering global warming. Energies 17:1–20. https://doi.org/10.3390/en17092017
9. Delgado-Gutierrez E, Canivell J, Bienvenido-Huertas D, Rubio-Bellido C (2022) Improvement options of a social housing prototype in different climate zones in Ecuador. Buildings 12(7):989
10. Vallejo-Coral EC, Vásquez-Aza F, Godoy-Vaca L et al (2023) Assessment of the thermal behavior in social housing in hot humid climate in Ecuador BT—trends in artificial intelligence and computer engineering. In: Gómez OS, Rosero Miranda R et al (eds) Botto-Tobar M. Springer Nature Switzerland, Cham, pp 442–454
11. Ministerio de Desarrollo Urbano y Vivienda (MIDUVI) (2018) Eficiencia energética en edificaciones residenciales (NEC-HS-EE)
12. Fundación Bariloche (2018) Consultancy for the survey and development of sustainable building performance standards for the Galápagos archipelago-RG-72384
13. Colmenero Fonseca F, Rodríguez Pérez R, Perlaza Rodríguez J, Palomino Bernal JF, Cárcel-Carrasco J (2024) Sustainable built environments: building information modeling, biomaterials, and regenerative practices in Mexico. Buildings 14(1):202
14. Vellei M, Herrera M, Fosas D, Natarajan S (2017) The influence of relative humidity on adaptive thermal comfort. Build Environ 124:171–185
15. Internacional Organization for Standardization (2007) ISO 10456:2007:Building materials and products—Hygrothermal properties—Tabulated design values and procedures for determining declared and design thermal values

16. Ministerio de Desarrollo Urbano y Vivienda (MIDUVI) (2018) NEC Norma Ecuatoriana de la Construcción. Miduvi pp 1–48
17. Gallardo A, Palme M, Beltrán RD et al. (2016) Analysis and optimization of the thermal performance of social housing construction materials in Ecuador. In: 32nd International conference on passive and low energy architecture cities, Buildings, People: Toregenerative Environments
18. Godoy-Vaca L, Vallejo-Coral EC, Martínez-Gómez J et al (2021) Predicted medium vote thermal comfort analysis applying energy simulations with phase change materials for very hot-humid climates in social housing in ecuador. Sustain Switz 13:1–31. https://doi.org/10.3390/su13031257
19. Instituto Nacional de Estadística y Censos—INEC (2015) Censo de Población y Vivienda Galápagos. 22
20. Steinitz-Kannan M, Lo C, Mari J (2020) History of limnology in Ecuador: a foundation for a growing field in the country. Hydrobiologia. https://doi.org/10.1007/s10750-020-04291-1
21. Fundación Charles Darwin (2023) Datazone. In: Base de datos climatológicos
22. NASA Prediction of Worldwide Energy Resource Project
23. Gomes R, Ferreira A, Azevedo L et al (2021) Retrofit measures evaluation considering thermal comfort using building energy simulation: two Lisbon households. Adv Build Energy Res 15:291–314. https://doi.org/10.1080/17512549.2018.1520646
24. Charles Darwin Foundation Galapagos (2023) Base de datos climatológicos. In: Puerto Ayora
25. ASHRAE Guideline 14–2014 (2014) Measurement of energy, Demand, and water savings. Atlanta, GA, USA
26. Baba FM, Ge H, Zmeureanu R, Wang LL (2022) Calibration of building model based on indoor temperature for overheating assessment using genetic algorithm: methodology, evaluation criteria, and case study. Build Environ. https://doi.org/10.1016/j.buildenv.2021.108518
27. Sun K, Specian M, Hong T (2020) Nexus of thermal resilience and energy efficiency in buildings : a case study of a nursing home. Build Environ 177:106842. https://doi.org/10.1016/j.buildenv.2020.106842
28. Litardo J, Macias-zambrano JA (2019) Measuring the effect of local commercial roofing samples on the thermal behavior of a social interest dwelling located in different climates in Ecuador IMECE2019–11472 measuring the effect of local commercial roofing samples on the. https://doi.org/10.1115/IMECE2019-11472
29. Vallejo EC, Vásquez-Aza F, Godoy-Vaca L et al. (2022) Assessment of the thermal behavior in social housing in hot humid climate in Ecuador. In: Trends in artificial intelligence and computer engineering. pp 442–454
30. Bhikhoo N, Hashemi A, Cruickshank H (2017) Improving thermal comfort of low-income housing in thailand through passive design strategies. Sustainability 9:1–23. https://doi.org/10.3390/su9081440
31. García Y, Cuadrado J, Blanco JM, Roji E (2018) Optimizing the indoor thermal behaviour of housing units in hot humid climates: Analysis and modelling of sustainable constructive alternatives. Indoor Built Environ 28:772–789. https://doi.org/10.1177/1420326X18793965
32. Yuliani S, Hardiman G, Setyowati E, Setyaningsih W, Winarto Y (2021) Thermal behaviour of concrete and corrugated zinc green roofs on low-rise housing in the humid tropics. Architect Sci Rev 64(3):247–261. https://doi.org/10.1080/00038628.2020.1751054
33. Vallejo-Coral EC, Vásquez-Aza F, Godoy-Vaca L et al (2023) Assessment of the thermal behavior in social housing in hot humid climate in Ecuador. Lect Notes Netw Syst. https://doi.org/10.1007/978-3-031-25942-5_35
34. Kong D, Liu H, Wu Y et al (2019) Effects of indoor humidity on building occupants thermal comfort and evidence in terms of climate adaptation. Build Environ 155:298–307. https://doi.org/10.1016/j.buildenv.2019.02.039
35. ASHRAE (2017) ASHRAE Handbook: Fundamentals

Sustainable Solutions and Climate Adaptation

Sustainable Solutions for Urban Flooding: The Case Study of Alcala de Guadaíra (Andalusia, Spain)

Laura Esposito, Inmaculada Guzmán Carrizosa, Miguel León Muñoz, Marta Torres González, and Francisco Javier Guevara Garcia

Abstract This chapter addresses the critical issue of soil waterproofing, often overlooked despite its significant environmental impact. Soil waterproofing, primarily caused by sealing materials like asphalt and concrete, disrupts natural soil functions, exacerbating land degradation and water consumption. The study focuses on the city of Alcala de Guadaíra, where 90% waterproof soil has led to numerous problems since 1900, including floods and landslides. The research proposes solutions at both private and public levels, emphasizing sustainable urban drainage systems (SUDS) and private building interventions. These solutions involve intermediate storage on private properties, functioning as energy rooftops utilizing water as an energy source. The findings underscore the potential of public–private cooperation in managing urban rain, highlighting its effectiveness in preventing and mitigating floods.

1 Introduction

'*The past is a safe haven. The past is a constant temptation. And yet, the future is the only place we have to go*' [1]. Technicians, public administrations, and entrepreneurs have long adhered to outdated urban development practices. These habits have caused tangible harm, prompting a slow and uneven shift toward new construction paradigms worldwide.

In construction, discussions often focus on thermal issues, solar radiation, or waterproofing measures, while overlooking the causes of soil impermeability. In

L. Esposito
Autostrade per l'Italia SPA, Genova, Italy

I. Guzmán Carrizosa
Magnun Capital, 41010 Sevilla, Spain

M. León Muñoz · M. Torres González · F. J. Guevara Garcia (✉)
Department of Building Construction II, Universidad de Sevilla, Sevilla, Spain
e-mail: guevara@us.es

M. Torres González and C. Rubio Bellido (eds.), *Towards Low and Positive Energy Buildings*, Green Energy and Technology, https://doi.org/10.1007/978-3-031-70851-0_9

countries prioritizing sustainability, building large structures on the ground is sometimes seen as an opportunity to mitigate the greenhouse effect. For instance, the California Academy of Sciences in San Francisco, designed by Renzo Piano, is acclaimed as the most sustainable museum in the world due to its 55,000 photovoltaic cells and large garden roof, providing efficient thermal insulation. This building exemplifies 'Green Building' or Nearly Zero Energy Building (NZEB), characterized by low environmental impact and the use of alternative materials and energy sources.

Modern buildings must be self-sustaining and environmentally neutral throughout their life cycles. Adhering to sustainable construction principles breaks the traditional link between building activities and waste, pollution, and land use. 'Green Buildings' offer numerous benefits, including a 30% decrease in energy consumption, a 27% increase in occupant satisfaction, a 35% reduction in emissions, a 19% reduction in maintenance costs, and a 70% decrease in waste [2].

Conversely, soil waterproofing has several negative impacts on water, biodiversity, climate, urban air quality, food security, and human well-being. These effects are detailed in the European guidelines 'Guidelines on Best Practices to Limit, Mitigate, or Compensate for Soil Sealing,' which provide solutions for problematic areas like Alcalá de Guadaíra [3].

Many cities with consolidated urban centers face significant problems due to a lack of natural resource protection. Cities like Alcalá de Guadaíra, Mexico City, and Venice experience significant damage from rainfall. Soil waterproofing alters water basins' environmental status, disrupting ecosystems and water services. Global warming exacerbates water scarcity, with reduced wetlands and permeable soils increasing flood risks. Waterproofing prevents soil from absorbing rainwater, leading to increased surface runoff, flooding, and surface pollution.

The three main direct impacts on water due to increased soil waterproofing are: (i) reduced water infiltration rates (both shallow and deep), significantly limiting open space areas, (ii) decreased time for infiltration on slopes, increasing surface runoff, which can lead to flooding and surface pollution, and (iii) reduced air-cooling evapotranspiration in urban areas [4]. These impacts exacerbate environmental issues and highlight the need for sustainable urban planning and soil protection measures.

Alcalá de Guadaíra suffers from these issues, with surface water unable to infiltrate back into the aquifer, overwhelming the urban drainage system. In sloped areas, high-energy runoff causes significant damage. The 'Green SUDS' project aims to address these challenges with sustainable urban drainage systems. This project includes design improvements to traditional collection systems and evaluates new sustainable drainage techniques, integrating hydrological solutions, local regulations, urban design, and long-term sustainability.

Technicians and professionals from EMASESA,[1] the municipal water company, support this project, facilitating diverse perspectives and services. The project aligns with European guidelines, promoting integrated and sustainable urban design strategies essential for developing Smart Cities [5]. The aim is to complement studies

[1] The supply and sustainable management of water are ensured by EMASESA, the local water enterprise in the city, guaranteeing quality of life and environmental preservation.

on the efficiency and viability of SUDS solutions in Alcalá de Guadaíra, assessing unique solutions for large residential buildings.

To manage inevitable transformations, sources of expenditure, and pollution, the project will conduct calculations for a limited area, selecting targeted interventions to mitigate rainwater's negative effects. An in-depth study of lithography and critical rainfall events will provide valuable data to simulate storm events and evaluate the contributions of proposed solutions. The specific objectives are to identify urban drainage solutions applicable to any city, develop water storage solutions to reduce peak flows during storms, and minimize the need for stormwater tanks.

Soil impermeability is a relevant issue across Europe, prompting various proposals to mitigate its effects. European guidelines, such as those presented at the 2012 'Soil Remediation and Soil Sealing' conference in Brussels, aim to prevent land network absorption by 2050 [6]. The European Commission's 2011 report 'Guidelines on Best Practices to Limit, Mitigate, or Compensate for Soil Sealing' compiles examples of effective practices across the EU [7]. These guidelines target authorities, spatial planning professionals, and stakeholders, promoting sustainable land use practices and minimizing soil sealing's negative impacts.

Sustainability protocols, developed since the 1990s, evaluate environmental, ecological, and social aspects within the construction sector. These protocols aim to minimize environmental impact, enhance energy efficiency, consider buildings' life cycles, and improve indoor environmental quality. Typically structured around rating systems, these protocols assign scores based on specific criteria or 'credits.'

The Italian ITACA protocol (Fig. 1), suitable for evaluating new construction and recovery projects, measures environmental performance through parameters, indicators, and weights. Approved by the Apulia Region, it comprises five evaluation areas with criteria grouped into categories. Buildings earn points based on their adherence to these criteria, resulting in an overall sustainability score ranging from − 1 to 5.

This study focuses on three relevant criteria related to climate change, heavy rainfall, and resource conservation:

Fig. 1 Map of protocols used around the world [8]

- C.4.1: Management of gray water discharge
- PR.5: Collection and storage of rainwater
- C.4.3: Promotion of soil permeability.

These criteria aim to reduce wastewater discharge, encourage rainwater harvesting, and minimize disruption and pollution of natural waterways. Increased urbanization and waterproofing have led to significant health and environmental challenges, including surface pollution, groundwater depletion, air pollution, heat islands, flooding, and the proliferation of concrete and asphalt surfaces.

Stormwater storage systems, as part of the 'WATER AS ENERGY' initiative, facilitate building air conditioning without conventional systems. This involves a centralized ventilation system with enthalpy recuperators and a water battery to harness renewable energy sources like geothermal energy, energy roofs, and solar energy from swimming pools.

This research draws on the concept of adaptive comfort as elucidated by Pérez-Fargallo et al. [9]. Studies on adaptive comfort demonstrate that the range of thermal comfort is considerably broader when air conditioning is not used, typically ranging between 20 and 27 °C, compared to the narrower comfort range achieved with air conditioning, which seldom deviates more than ± 1 °C from the desired temperature [10]. Comparative studies on energy demand utilizing the adaptive thermal comfort approach have been conducted in various contexts, including office buildings in Spain [11], and residential buildings in dry and warm climates by Escandón et al. [12].

Particularly noteworthy are studies focusing on the application of adaptive comfort principles in social housing, such as those by Rubio-Bellido et al. [13]. Furthermore, research within this line aims to assess the impact of occupancy patterns on energy demand in residential buildings through the implementation and monitoring of various systems. This includes studies by Cuerda Barcaiztegui et al. [14] and Escandón et al. [12], which analyze presence patterns and energy demand using measured data and simulation models. Some studies even employ machine learning algorithms to predict clothing insulation [15].

For the utilization of water as an energy source for urban temperature control, thermal mapping of target areas can be conducted, as demonstrated by Romero Rodríguez et al. [16].

2 The Case Study: Alcalá de Guadaíra (Seville, Spain)

Rapid urban expansion in many cities worldwide, including Alcalá de Guadaíra, has led to the neglect of critical factors like proper drainage system design and effective rainwater evacuation strategies. This oversight, often termed "discharge of responsibility," is evident in the placement of scuppers solely along road edges, far from newly developed areas. Although these placements may comply with certain urban planning standards, they do not reflect the city's growth dynamics, marked by a lack of permeable surfaces and significant slopes.

The primary factors contributing to flooding incidents in Alcalá are the absence of permeable surfaces and steep gradients. Surface water accumulates due to poor infiltration, and its velocity prevents it from entering scuppers. As a result, water flows along impermeable streets in the city center, increasing flood risks. In fact, data from the Insurance Compensation Consortium (hereinafter, CCS) [17], reveals that from 1987 to 2004, floods accounted for 56.91% of reported damages in Spain, followed by atypical temperature events at 29.65%, terrorism at 5.50%, and other causes at 7.94%.

Many Spanish cities have suffered significant flood damage. According to the CCS, of the 35 most significant disasters recorded from 1971 to 2004, 31 were due to floods. Urban environments are the primary sources of flooding, mainly associated with short-term, intense rainfall events. Several factors limit effective mitigation measures:

- A comprehensive understanding of flooding, including historical data and future predictions, is essential. However, accurately predicting the location, timing, and intensity of rainfall events remains challenging.
- Climate change may cause variability in precipitation patterns over time, leading to uncertainties in quantifying these changes [18].
- Urban development trends, driven by global population growth, have led to densely populated areas and urban sprawl, resulting in increased land waterproofing.
- The transition from permeable to impermeable materials in basin coverings accelerates surface water runoff, increases peak flow rates, reduces peak flow durations, and deteriorates water quality.

Thus, changes in the hydrological response of the basin are illustrated in Fig. 2. The graph shows that the progressive waterproofing of urban soils and the use of concrete and asphalt significantly increase the runoff coefficient. This leads to an earlier and substantially higher peak flow rate.

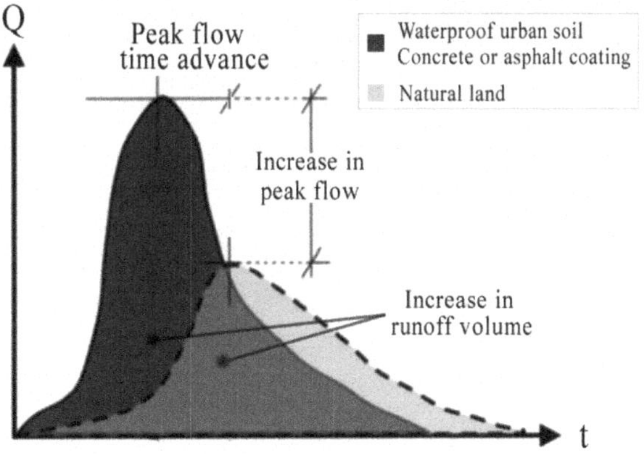

Fig. 2 Hydrogram basin response

Once the curve for extreme storm episodes is calculated for the city, the study aims to redistribute the peak flow by advancing it in time to prevent damage to the urban network. This requires the installation of both private and public collection systems. The general problem of rainwater drainage in urban basins can be subdivided into four sub-problems [18]: (i) determining the volume of water to be evacuated, (ii) establishing the water discharge system into the network, (iii) designing discharge conduction systems, and (iv) implementing discharge systems to the environment.

The exponential growth of cities and urbanized areas has led to the proliferation of impermeable land, where water cannot adequately drain and complete its natural cycle. Before urbanization, these areas could naturally manage the water cycle. However, the reduction of green and permeable spaces has diminished natural interception, evapotranspiration, and infiltration. Consequently, significantly higher volumes of runoff water are generated, and response times are accelerated, increasing the risk of flooding.

Recent rainfall events have caused severe damage and issues in the main streets and avenues of Alcalá de Guadaíra's city center, with infrastructure appearing overwhelmed. Surface runoff from rainfall exacerbates environmental degradation by carrying garbage and pollutants, leading to the saturation of the sewage system and posing serious hazards to road traffic. It also impedes the free movement of people and allows water to infiltrate homes, causing further damage. In response, sustainable urban drainage initiatives aim to enhance water management and collection quality, mitigate flooding, and promote sustainable urban development.

Despite commendable efforts to enhance the traditional urban drainage system in Alcalá de Guadaíra, further interventions are imperative. For instance, the capacity of the pipe and tank system has been increased, and the surface catchment area expanded by deploying scuppers along the width of roads.

Figure 3 shows two characteristic streets in the heart of Alcalá de Guadaíra. The left photograph depicts a steep, fully impermeable street with a single water collection point at its center, typical of the area. The right photograph illustrates a road prone to flooding during heavy rainfall, exacerbated by numerous natural channels that gather water and a network of reinforced concrete gutters.

Implementing sustainable drainage systems like SUDS or storage tanks for homes or individual plots[2] in Alcalá de Guadaíra requires various tests on the prevalent *albero soil*,[3] both in the laboratory and on-site:

- *Infiltration Tests*: Using methods such as concentric rings or the Müntz method to assess water infiltration rates, providing data for water balance calculations to determine soil moisture and potential recharge capacity.

---

[2] Stream of water that is poured when passing its natural and artificial reservoir (or channel).

[3] The soil in this area is mainly a pure earth mixture, with compositions that vary depending on specific locations. It resembles clay, showing a red color and low permeability ranging from 10^{-5} to 10^{-9}. These characteristics have historically made it useful for ancient street construction because its waterproof nature prevents vegetation growth, keeping surfaces clean. Moreover, its low tendency to expand makes it stable and suitable for building foundations.

Fig. 3 Streets in the center of Alcala (left) and grids system added (right)

- *Seasonal Monitoring*: Tracking seasonal fluctuations in the water table to understand groundwater flow dynamics.
- *Transmissivity Calculations*: Determining transmissivity values to characterize the aquifer, considering the heterogeneity of conductivity values, particularly increasing toward discharge areas.
- *Surveys*: Conducting surveys to objectively determine the depth of multiple soil layers and geological units, offering valuable insights into soil composition and structure.

The upper basin of Alcalá de Guadaíra, according to the lithological classification by the Junta de Andalucía, consists of impermeable layers such as marls, clays, calcarenites, and limestones. Despite efforts to apply sustainability and permeability principles, many challenges persist due to the terrain's characteristics. Addressing these issues in such a challenging context can serve as a blueprint for similar challenges in other cities worldwide. The impermeability in Alcalá de Guadaíra is influenced by both human activities and inherent lithophysical features.

Sustainable urban drainage systems (SUDS) designed to enhance permeability must adapt to the unique shape and subsoil conditions of Alcalá de Guadaíra. They are ideally installed near river areas where soil types like sand, silts, gravels, and clays, as identified in the lithographic plan by the Junta de Andalucía (SIGMA), exhibit greater permeability to water. The primary objective of SUDS is to replicate the natural hydrological cycle as closely as possible, compensating for the low permeability of the albero soil. SUDS aim to minimize the impacts of urban development on runoff quantity and quality, enhance landscape integration, and increase the social and environmental value of urban interventions [19].

SUDS are designed to complement, not replace, conventional drainage networks. They offer a permeable, modular, and eco-efficient solution that captures water before it enters the network, allowing for its collection at the source, groundwater recharge, or discharge into uncontaminated channels [20]. Many cities worldwide have already implemented rainwater collection systems or infiltration strategies, adopting approaches known as '*sponge cities*' [20]. In Spain, while progress is being made in several cities, implementation remains slow and limited. However, there is a clear desire for improvement, supported by government legislation such as Royal Decree 638/2016, which mandates the inclusion of SUDS in new urban developments.

In addition to SUDS, collective rainwater collection systems in urban areas and individual plots play a crucial role in mitigating flooding issues, as recommended by sustainability protocols. In cities like Alcalá de Guadaíra, where significant surface areas are waterproofed by the private sector, SUDS systems alone may not suffice, especially in densely populated areas. Individual devices are necessary to effectively serve one or more plots.

3 Methodology

3.1 Analysis of the Case Study Characteristics

In order to enhance the quality of the soil in residential areas, it is imperative to equip buildings with tanks and systems designed to capture and reuse rainwater and/or gray water. Outdoor spaces should be permeable to allow water infiltration into the groundwater table. New buildings should not only avoid exacerbating soil degradation but also contribute to its sustainability. Green roofs can help offset the impermeability caused by traditional waterproofing methods.

To assess the quality and composition of the soil in Alcala, the QGis software has been utilized. Cartographic data play a crucial role in understanding the mineralogical composition of the soil, which significantly influences the feasibility and functionality of newly implemented SUDS devices (Fig. 4). The region is marked by an impermeable soil predominantly composed of calcarenite. In proximity to the river, however, permeability marginally increases, with the soil primarily consisting of sand and silt. The specific slopes of the streets within the historic center were analyzed (Fig. 5), revealing a notable buildup of rainwater during heavy precipitation events. Consequently, these unpredictable weather occurrences lead to an overload of the urban water disposal network, posing a risk of network component failure.

Following the guidelines of the ITACA protocol at the urban scale, the passages made to calculate the permeability of the soil of the area composed of buildings and streets of greater risk of flooding are: (i) identification of permeable areas and creation of polygons in QGIS, (ii) calculation of the area of each polygon, (iii) sum

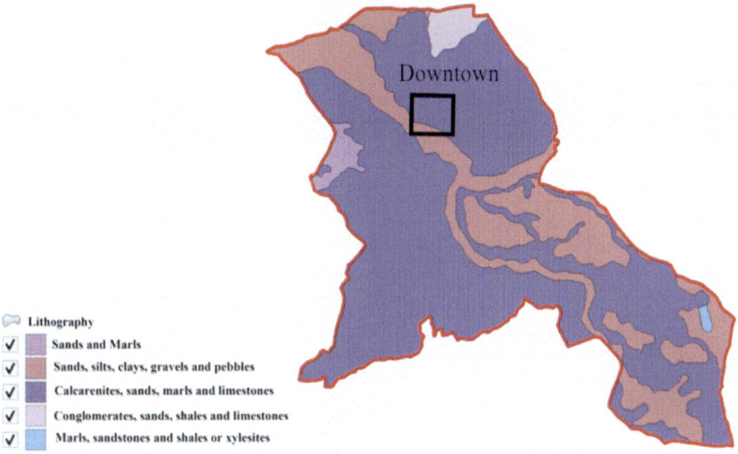

Fig. 4 Lithography of Alcala de Guadaíra extracted from the QGIS

Fig. 5 Alcala de Guadaíra: Slope of the streets (from 10 to 29%) extracted from QGIS

of permeable areas, (iv) attribution of permeability indices, and (v) calculation of the permeability index of the area.

The total area of green spaces (Fig. 6), each of which is assigned a permeability index of 1, amounts to 123,183.84 m^2 and the area of the basin of influence in the *Nuestra Señora del Águila St.* is 2,582,363.00 m^2. Thus, the permeability index will be (Eq. 1):

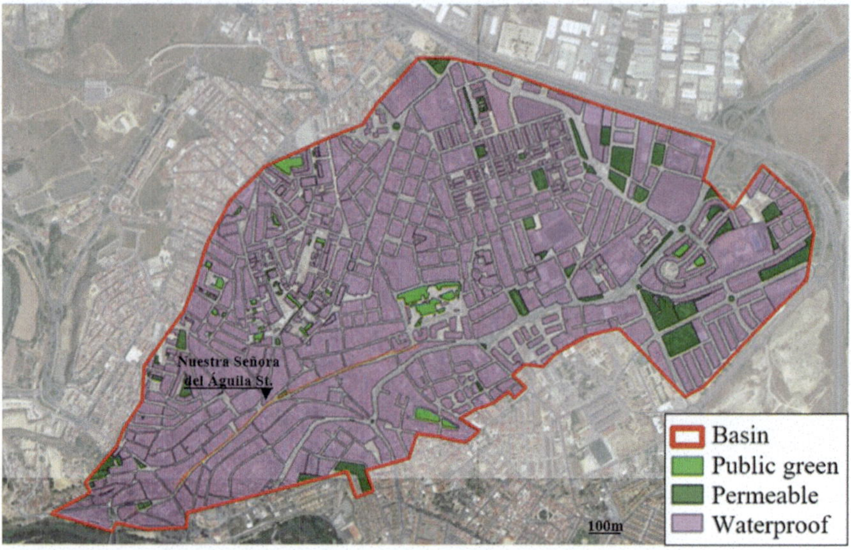

Fig. 6 Basin of influence in the street Nuestra Señora del Águila

$$\text{Permeability index} = \frac{123{,}183.84}{2{,}582{,}363.00} * 100 = 4.77\% \qquad (1)$$

Following comprehensive investigations encompassing standards, geological assessments, on-site city configuration analyses, QGIS cartographies, and the calculation of permeability indices according to the ITACA protocol, historical precipitation data spanning from January 1, 1920, to January 1, 2020, were acquired from the freely accessible historical database of Andalusia. These data were sourced from the meteorological station situated at Seville Airport, the closest station to Alcala. The objective was to identify and understand the rainfall events responsible for the most significant damage in the area.

Using data from rainfall events that resulted in damage [21], an estimation of the flood occurrences on October 2, 2007, December 12, 2008, and September 14, 2018, was derived by comparing hourly data from these dates. Additionally, data from October 22, 2016 (Fig. 7), were pertinent, with the hourly breakdown as follows:

Upon comparison of daily data from the same period, it was observed that the aforementioned day recorded the highest precipitation level, reaching 105.20l/m^2. This value closely approached the maximum precipitation recorded within the last century, which amounted to 109.30l/m^2 (Fig. 8).

Given the limited accuracy of Internet-sourced data for simulating short-duration water pump issues lasting a few hours, assistance has been sought from the EMASESA organization. This organization possesses a significant database containing information on rain events recorded in 10-min intervals. The rain event under examination aligns with the scenario outlined in the guide for sewer network

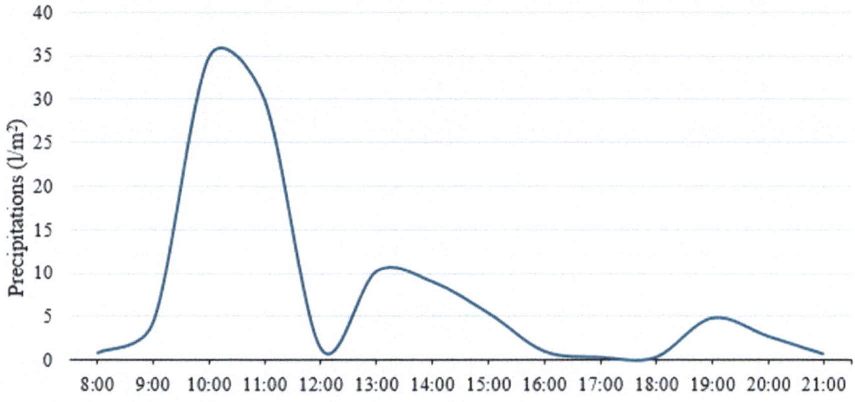

Fig. 7 Rainfall data 22/10/2016 in Alcalá de Guadaíra [21]

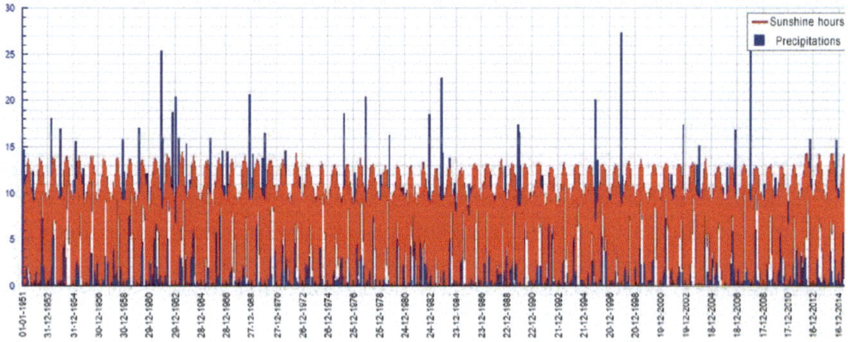

Fig. 8 Representation of the maximum recorded values of rainfall and hours of sunshine of the last 100 years in Seville [22]

characterization, recurring every five years. While this event may not exhibit excessive intensity, it nonetheless results in flooding across Andalusian cities. Most of the projection aligns with the limit event depicted in Fig. 9.

To know the effective functionality of solutions for the reduction of flood risk, two different models were adopted for two solutions that can be implemented in private construction: **collection cisterns and green roofs** [24].

The calculation begins with the determination of the flow filling at a certain point and the established return period, of Eq. 2 [23]:

$$Q = \frac{C \cdot I \cdot A}{0.36} \tag{2}$$

where Q (l/s) is the project rainwater flow or peak flow; C is the average flow coefficient (between 0.0 and 1.0) whose value could be 0.95 for large paved areas

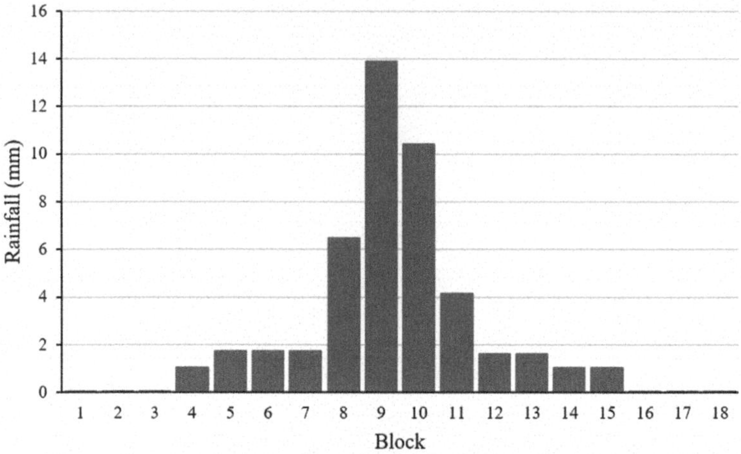

Fig. 9 Project rain event, histogram values for blocks of 10 min, and total duration of 3 h. *Source* [23]

(large car parks and large spaces without gardens), 0.85 for urban areas (roads, squares and high-rise buildings), 0.50 for residential areas (urbanizations in which the single-family building is mixed with gardens), and 0.20 for unpaved areas (parks and gardens); I (mm/h) is the intensity of the rain that corresponds to the maximum event for a specific period of return and duration equal to the concentration time (T_c); and A (ha) denotes the surface area subjected to the rainfall.

This method enables the calculation of the accumulated flow in a typically water-proofed roof (with a flow coefficient of 0.95) that ultimately reaches the public network. Subsequently, this accumulated flow will be contrasted with the flow at the outlet of the meteoric water accumulation tank, ideally situated to serve the same surface area utilized for computing the outflows from the green roof. Through this comparison, the results obtained can quantify the mitigation achieved by the collection system, which may vary in form and material.

There will be tanks (buried or on the surface), roofs, courtyards, etc. Each tank will have an overflow to protect the maximum filling level, capable of evacuating all the collected water at any given time. In some cases, a drain can be installed that can evacuate a permanent but limited flow to the network until it is filled, where the overflow is located.

For the size of the tank, the calculation of the Italian standard UNI/TS 11445:20121 [25] was taken as a reference, according to which it is necessary to consider as optimal volume 6% of the minimum between:

- The contribution of rainwater was determined concerning the intensity of precipitation, the receiving surface (for sloping roofs, only the horizontal projection is considered), and the flow coefficient.
- Need for service water, depending on the type of user, the number of users, and the specificity of the services of use requested.

In the case of green roofs, it was considered that the collected water will first fill the volume of soil, then the nearby collection points will be discharged into the drain, and finally the distant points will be drained, which may take minutes to reach, greatly reducing the runoff coefficient in both cases and, consequently, greatly delaying the exit of the water. In this chapter, an area of 355 m^2 was considered for the green roof simulation. The roof surface is the top floor of a flat, and the service exclusively aims to replace the use of drinking water with rainwater solely for toilet purposes. Considering an average apartment of 100 m^2 composed of two double rooms, the number of users is 4 per 100 m^2, multiplied by 2 since the majority of buildings in the center of Alcala de Guadaíra have two floors. Additionally, a daily water consumption of 35 l per user was set.

For the calculation of the ideal volume of accumulation, the data of the average annual rain fall is equal to 550 mm/year, according to the information offered by the Spanish Government [22].

The minimum between the volumes calculated concerning precipitation and need will be multiplied by 0.06 (rainfall yield per year) and by 1.5 (a safety coefficient that allows to obtain a good efficiency of the system also in the presence of important variations of local rainfall and the modalities of use of the water), resulting in the theoretical volume of the recoverable rainwater tank per year (Eq. 3).

$$V_{\text{optimol}} = 0.06 \cdot \min\left[F; V_{\text{piogg}}\right] \cdot 1.5 \tag{3}$$

$$\text{Ratio} = V_{\text{cist}}/V_{\text{ott}} \cdot 100 \tag{4}$$

The volume of the cistern $V_{\text{minCISTERN}}$ destined to the collection of meteoric waters will be chosen among the commercial volumes and analyzed the results during the pluvial events (daily data recorded by SAIH [21]) in the period from 1/1/2008 to 08/05/2020, considering various hypothetical dimensions of the tank.

Finally, the percentage ratio between the volume of recoverable rainwater and the volume actually recovered (Eq. 4) could be calculated and thus calculate the performance indicator, a parameter used to examine the level of performance of the building with respect to the evaluation criteria of the ITACA protocol criterion sheet rp.5.

3.2 *Simulation of the Model*

Previous works on the topic deal with an experimental monitoring field carried out on the green roof of the University of Genova in 2011 from which it is possible to know the simple conceptual model (with careful calibration and validity) that have the potential to provide sufficiently valid estimates of useful parameters to evaluate the hydrological performance of the green roof [26]. The structure of the conceptual model is represented in Fig. 10.

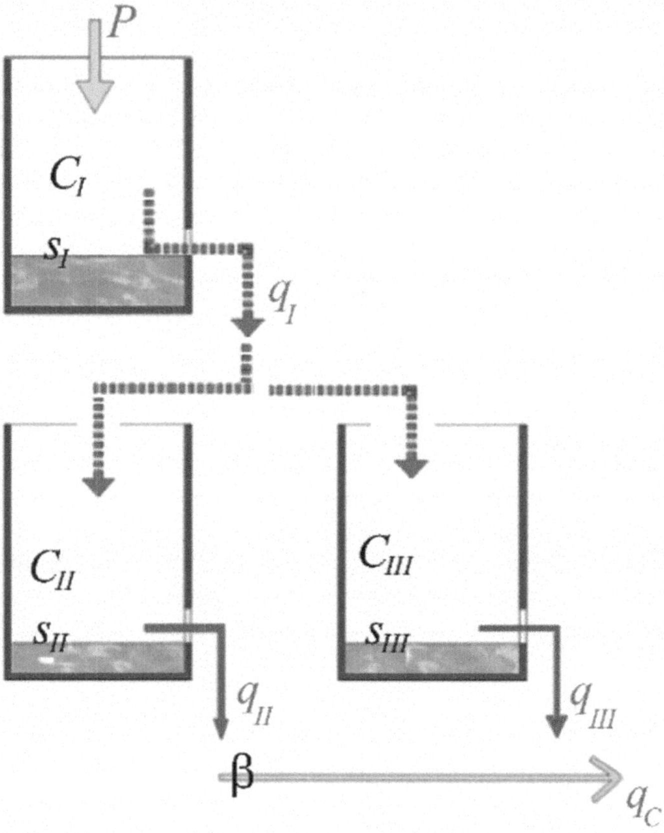

Fig. 10 Structure of the conceptual model. *Source* [26]

The green roof system is simulated using a cascade of three linear deposits: the first represents the process of infiltration in topsoil, while the second and third describe lateral flow. In particular, the second deposit corresponds to the roof area away from the outlet sections, it is the slow response area, in which there is a flow delay in the subsoil. The third deposit corresponds to the roof area above the drainage channel, a rapid response area characterized by spontaneous and rapid flow.

The behavior of the first deposit is described by the pair of equations (Eqs. 5 and 6) [26], where the subscripts indicate the deposit number; C is the reservoir constant, [T^{-1}]; s is a threshold value [L]; q is the specific tributary per unit area [LT^{-1}]; h is the water depth [L], and P is the precipitation [LT^{-1}].

$$\begin{cases} q_I^i = 0, & \text{if } h_I^i < s_I \\ q_I^i = C_{(I)} \cdot (h_I^i - s_I), & \text{if } h_I^i \geq s_I \end{cases} \tag{5}$$

$$h_I^i = P^i + h_I^{i-1} - q_I^{i-1} \tag{6}$$

The behavior of the second deposit is described by the pair of Eqs. 7 and 8 [26]:

$$\begin{cases} q_{II}^{i} = 0, & \text{if } h_{II}^{i} < s_{II} \\ q_{II}^{i} = C_{II} \cdot \left(h_{II}^{i} - s_{II}\right), & \text{if } h_{II}^{i} \geq s_{II} \end{cases} \tag{7}$$

$$h_{II}^{i} = q_{I}^{i} + h_{II}^{i-1} - q_{II}^{i-1} \tag{8}$$

Analogous relationships are valid for the third deposit. In particular, for the three deposits, the constants, respectively, represent the time used by water to move vertically toward the lower impermeable stratum (CI) and to move horizontally toward the exit sections of the slow response (IIC) and fast (CIII) response areas.

The specific flow of the green roof is calculated as a linear combination of distribution of the second and third deposits through the weighting coefficient, as follows (Eq. 9) [26]:

$$q_{C}^{i} = \beta \cdot q_{II}^{i} + (1 - \beta) q_{III}^{i} \tag{9}$$

Therefore, the weighting coefficient allows to divide the ceiling into areas of slow and fast response (Table 1).

With a calibrated and validated model available in the experimentally collected database, it becomes possible to evaluate the hydrological performance of the green roof in response to rain events in Alcalá de Guadaíra. The dimensions and characteristics of the roof used by Palla et al. [26] were maintained (Fig. 11). The roof under consideration belongs to the category of intensive green roofs, characterized by a thickness of 40 cm (see Table 2 for differences among green roof typologies).

The hypothesis of the extensive roof was also simulated, in which the threshold heights are cut in halve ($s1$ from 41 to 21 mm, and $s2/3$ from 2 to 1 mm) and the deposit constants are doubled ($C1$ from 0.0313 to 0.0626, $C2$ from 0.0031 to 0.0062, and $C3$ from 0.03 to 0.06), since a lower thickness of the filling states of the roof will correspond to a higher speed of displacement of the water to its interior, that is, the time used by the water to move vertically to the impermeable stratum and horizontally toward the exit sections, will be lower.

Multiple analyses have been completed, providing insights into the challenges associated with land permeability in various cities. A comprehensive list of diverse rainwater drainage systems tailored to different city landscapes and construction types has been compiled based on their respective characteristics. While the proposed calculation models could potentially apply to other roof types or green roofs,

Table 1 Parameters of the conceptual model of the green roof [26]

	C (min^{-1})	s (mm)	B (–)
Reservoir I	0·0313	41	–
Reservoir II	0·0031	2	0·24
Reservoir III	0·0300	2	0·24

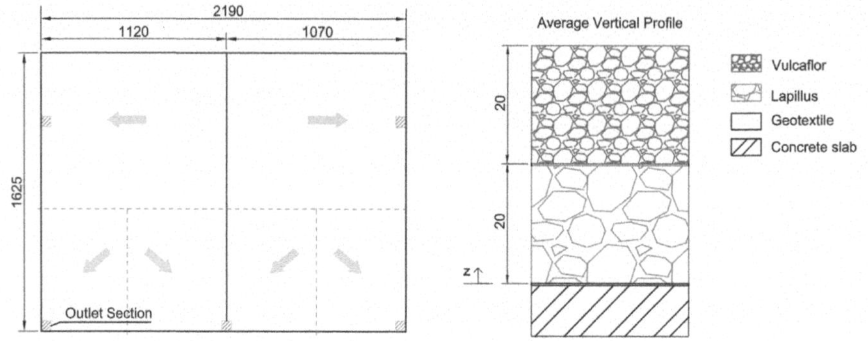

Fig. 11 Plan view of the green roof experimental site and outline of the average vertical profile (dimensioned in cm). Retrieved from [26]

Table 2 Differences between green roof typologies

Intensive green roof	Extensive green roof
> 20 cm	< 20 cm
> 200 kg/m^2	60–150 kg/m^2
Accessible	Not accessible
Complex construction	Pretty easy to build
Complicated maintenance and need for drainage and irrigation systems	Easy maintenance and irrigation not always necessary
A lot of plant diversity	Low plant diversity
> 80 €/m^2	< 80 €/m^2

further calculations would be required to determine appropriate and properly sized parameters.

The results that can be obtained with the implantation of rainwater collection systems and green roofs should not be underestimated, in addition to a time lag of the minimum peak of 2 h obtainable with this type of solutions, the percentages of flow reduction with respect to the maximum values are summarized in Table 3.

To complete the work, the analysis developed in the urban area, a simplified evaluation was developed, without taking into account the transfer time of the liquid mass, from the time the water drop reaches the roof, until it reaches the urban sewage network, which will depend on the material of the pipes, geometric configuration, etc., that depends on many parameters, but the objective of this chapter is exclusively qualitative and quantitative of the first level, approximately.

The hypothesis aims to analyze the phenomenon by linearly combining the effects of flow attenuation from various solutions, weighted according to the surfaces defined in QGIS. As flow rates vary while maintaining a constant final volume, we quantify this volume by multiplying the flow rate by the time interval.

Table 3 Reduction values obtained

Solution	Peak reduction (%)	Reduced flow (%)
Intensive green roof	100	95
Intensive saturated green roof	93	8
Extensive green roof	92	49
Extensive green roof saturated	86	4
Tank 15,000 L	40	70
Tank 25,000 L	46	77
Tank 35,000 L	46	81

Different scenarios have been compared through an analytical evaluation of the volumes in the assumption of a variable percentage of green cover with respect to the total occupied, which varies in the different cases is the accumulated volume, composed of a proportion of water from the green roofs and the rest of the traditional roofs.

Various meteorological scenarios (dry or saturated green roof), architectural (intensive and extensive green roof), and the percentage of green roof of the buildings (100, 50, and 25%) were compared. The surfaces calculated in GIS are presented in Table 4.

where the influx coefficient is the ratio of net rainfall corresponding to rainfall runoff and gross influx into the basin, related to the same rainfall event; the flow coefficient is the ratio of surface runoff measured in the closure section (including groundwater intake) and inflow to the basin, over the same period of time; and the permeability coefficient represents the ability of a surface to be traversed by fluids.

The sum of the surfaces of the flat roofs, in which green roofs of intensive and extensive type could be installed, will be considered for the calculation of comparison between scenarios.

The sum of the volume of water accumulated during the rain event (46.63 mm/ m^2) is equal to 85,916,592 l, in all traditional roofs; with respect to this value, the percentages of improvement will be calculated, considering the sums of the volumes that leave the surfaces of the roofs in a period of time of 3 days starting from the critical event of rain.

Table 4 Respective surfaces and coefficients

Surfaces	Sum (m^2)	Influx coef.	Flow coef.	Permeability coef.
Tile roof	92,018.76	0.8	0.85–1	0
Flat roof	1,939,492.14	0.7	0.85–1	0
Wooded greens or grass	123,183.84	–	0.1–0.15	1
Land area	12,928.41	–	0.5–0.6	0.3
Waterproof to earth	414,739.86	0.7	0.85–1	0

4 Discussion of Results

By filtering the data in l/m^2, referring to the sum of the volume of rain accumulated in a day, and taking into consideration only the days in which the critical value of 46.63 l/m^2 was exceeded—sum of the rainfall recorded in the 3 h in which the limit event provided by EMASESA—it was possible to compare different volumes of accumulation and to value the most suitable not only according to the indications of the norm, but also according to the frequency of arrival of those that are too full.

The volumes compared are trade volumes: 15,000–25,000–35,000 l. The first one is closest to the optimal volume calculated according to the UNI/TS 11445:2012 standard and the last of them, is closest to the ideal volume of need, the second is an intermediate value.

Looking at Table 5, it is deduced that the volume that contains the most and decreases the flow at the outlet (1/m^2) is that of the third tank, the largest. The adoption of that volume could be inefficient due to the long dry periods that characterize the Mediterranean climate and due to the greater weight and volume that would load in the installation area. Remember that in Spain, rainfall does not occur throughout the year. Most of the rainfalls were focused on the months of October, November, April, and May, being almost non-existent in the summer months, so the tank must be oversized and install a purification system that keeps the water in conditions in those months in which there is a request, but it does not rain.

As can be seen in the graph (Fig. 12), flows in output of the three types of deposits and that of the cover without deposit (in sky blue), the inferiority of the flows is deduced (on average the reduction of percentage of the flows in the critical episodes is for the three deposits, respectively, 43.10–67.24–76.63%) with the exception of the dates of December 2009 and 2010, in which rain events have been important for at least three consecutive days and the tank has remained too full.

With the only reference to critical events (in which the daily accumulation exceeds 46.63 l/m^2), the percentages of the achievement of the too full, for the three hypotheses of volume of the deposits, are, respectively, 100%, 60%, and 50%.

Finally, an analysis is fulfilled on the frequency with which the too full tank is reached during the total observation period (4601 days). The first deposit (15,000 l) corresponds to a frequency of 3.4%, the second (25,000 l) a frequency of 2.6%, and the last (35,000 l) 2.2% [26].

The results obtained with the application of the equations retrieved from Palla et al. work [26], were depicted in Figs. 13 and 14. Both combined graphs present on the abscissa axis the time in minutes, and on the axis of ordinate on the right the scale of the values referred to I (intensity of rain in mm/min), and on the left the scale of values referred to the flows Q.

Successively to the rain event, which achieves the maximum intensity at 81 min, the green roof can retain all the water inside up to one hundred and forty minutes and then release the maximum amount of water after 7 h (the maximum value corresponds to minute 508). According to the percentages, the reduction of the peak is 100% and the reduction of the flow is 95%.

Table 5 Comparison of three trading volumes on rainy days, highlighting critical accumulation milestones

Data	Feb. 2, 2008	Dec. 12, 2009	Dec. 23, 2009	Dec. 12, 2010	Sep. 28, 2011	Apr. 4, 2016	May 8, 2016	Oct. 22, 2016	Nov. 25, 2011	Nov. 29, 2011
l/m^2	48.75	80.20	57.00	51.30	62.02	50.50	56.60	105.20	52.60	60.60
L accumulated	17,349	29	20,285	18,256	22,072	17,972	20,143	37,438	18,719	21,566
Water tightness	980	980	980	980	980	980	980	980	980	980
Filling $status_1$ (l)	15,000	15,000	15,000	15,000	15,000	15,000	15,000	15,000	15,000	15,000
L_{out1}	1900	208,640	19,305	17,276	6092	1992	13,321	21,458	12,893	7702
Q_{c1}	5.34	58.63	54.25	48.55	17.12	5.60	37.43	60.30	36.23	21.64
Filling $status_2$ (l)	16,900	25,000	25,000	25,000	21,092	16,992	25,000	25,000	25,000	22,702
L_{out2}	0	10,864	19,305	17,276	0	0	5540	11,458	4726	0
Q_{c2}	0.00	30.53	54.25	48.55	0.00	0.00	15.57	32.20	13.28	0.00
Filling $status_3$ (l)	16,900	35,000	35,000	35,000	21,092	16,992	30,540	35,000	35,000	22,702
L_{out3}	0	864	19,305	17,276	0	0	0	1458	4726	0
Q_{c3}	0.00	0.00	54.25	48.55	0.00	0.00	0.00	4.10	13.28	0.00

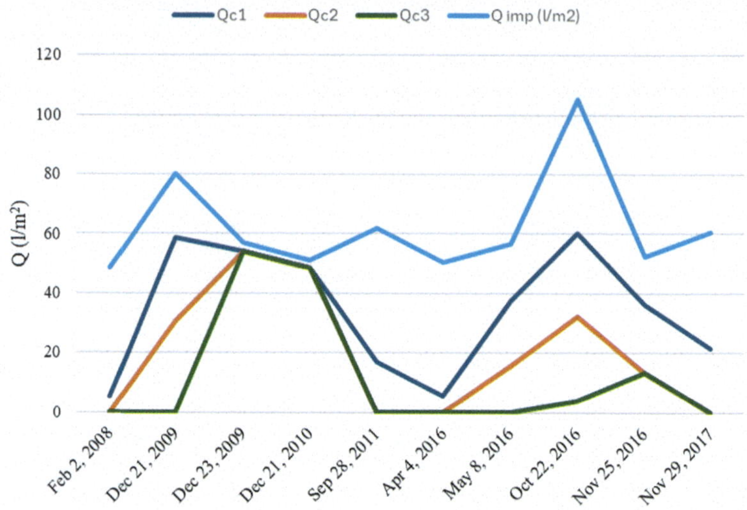

Fig. 12 Green roof drainage flow rate

Applying the formulas to an already saturated ceiling (height $h1$ at the beginning t equal to 0 is 41 mm), the results obtained are those represented in the following graphs.

In this case, the maximum flow at the exit is recorded at 222 min, so the time lag of the peak, with respect to the maximum of the rain event (81 min), is little more than 2 h; the reduction in flow is 8% and that of the peak is 93%.

The results obtained by simulating the extensive roof were summarized in Table 6.

The comparison between scenarios in the table is based on the total surface area of flat roofs suitable for installing intensive and extensive green roofs. The volume of water accumulated during the rain event (46.63 mm/m^2), totaling 85,916,592 l across all traditional roofs, serves as the baseline. Percentages of improvement were calculated relative to this volume, considering the cumulative volumes drained from roofs over a 3-day period following the critical rain event, as detailed in Table 7.

5 Conclusions

This chapter undertakes a comprehensive analysis to address permeability issues in urban areas. It includes an inventory of drainage systems suitable for various city contexts, contingent upon hygienic-sanitary requirements. The developed calculation model is adaptable to different types of green roofs.

Implementation of meteoric water collection systems and green roofs demonstrates significant benefits, achieving a 2-h delay in peak flow and reducing runoff by 81–95%. Additionally, a simplified evaluation assesses liquid mass transfer times to

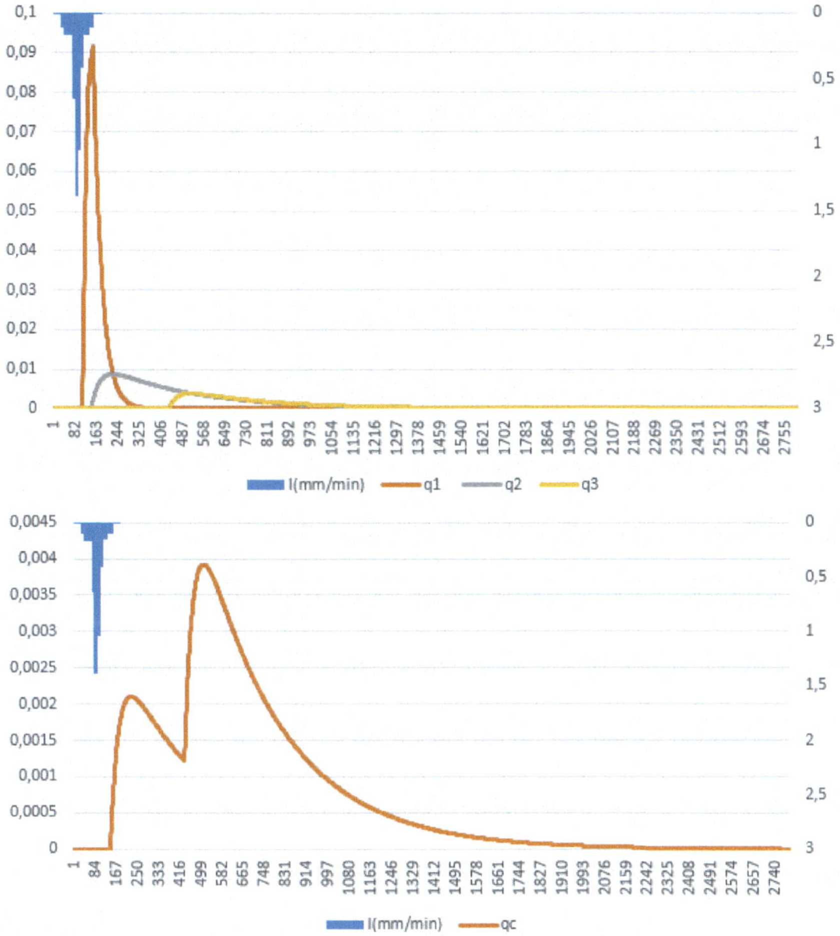

Fig. 13 Graphs of the results obtained by the dry roof spreadsheet

networks at a larger scale, highlighting potential benefits of proposed construction solutions.

To enhance soil quality and mitigate flood risks in Alcala and other cities, it is essential for residential buildings to incorporate rainwater and gray water retention systems. Permeable pavements and systems allowing water infiltration into aquifers are crucial. New buildings should aim for sustainability, integrating green roofs to counteract impermeability.

The analysis and modeling yield positive outcomes for urban hydrological risk management. However, achieving definitive urban-level results requires a more precise model accounting for specific hydraulic aspects of the urban network, such as pipe materials and roughness.

Fig. 14 Graphs of the results obtained by the saturated ceiling spreadsheet

Table 6 Summary of the results obtained

Ceiling status at $t = 0$	Peak reduction (%)	Reduced flow (%)	Peak lag (min)
Dry extensive roof	92	49	90
Saturated extensive roof	86	4	77

The study's initial findings in Alcala de Guadaíra underscore the importance of further examining the city's water network configuration. Detailed tests are necessary to precisely define soil parameters and simulate improvements using appropriate software, ultimately proposing effective solutions to the identified issues.

Table 7 Summary of the results for each scenario

Types of roof	Accumulation of volume on private land (l)			Percentage of reduction		
	100% green scenario	50% green scenario	25% green scenario	100% green scenario (%)	50% green scenario (%)	25% green scenario (%)
Intensive green roof	4090.60	45,003.60	65,460.10	95	48	24
Intensive saturated green roof	83,593.47	84,755.03	85,335.81	3	1	1
Extensive green roof	46,295.68	66,106.13	76,011.36	46	23	12
Extensive green roof saturated	87,025.01	86,470.80	86,193.70	0	0	0

References

1. Zabalbeascoa A, RodríguezMarcos J (1998) Renzo Piano. Gustavo Gili [etc.], Barcelona
2. Spain Green Builging Council (2023). http://www.spaingbc.org/files/LEED101Docs742ESPCorto.pdf. Accessed 3 June 2024
3. Commission E (2012) Guidelines on best practice to limit, mitigate or compensate soil sealing. In: Commission staff working document
4. Bruselas E, Consejo DEL (2015) ES
5. Rodriguez-Rojas M (2019) Sistemas urbanos de drenaje sostenible. Una oportunidad para la planificación de ciudades sensibles al agua. Revista de Obras Publicas 166:14–20
6. Prensa CDE (2012) Medio ambiente: Directrices para la limitación del sellado del suelo
7. COMISIÓN EUROPEA (2011) Hoja de ruta hacia una Europa eficiente en el uso de los recursos. Prensa comunidado 4:410–419
8. Perucca E, Molnar P, Francis R, Perona P (2010) Unravelling floods selection mechanisms for vegetation from laboratory experiments and implications for river restoration efficiency. In: EGU general assembly conference abstracts, p 5788
9. Pérez-Fargallo A, Rubio-Manzano C, Martínez-Rocamora A et al (2018) Linguistic descriptions of thermal comfort data for buildings: definition, implementation and evaluation. Build Simul 11:1095–1108. https://doi.org/10.1007/s12273-018-0455-7
10. Sánchez-García D, Bienvenido-Huertas D, Rubio-Bellido C, Rupp RF (2024) Assessing the energy saving potential of using adaptive setpoint temperatures: the case study of a regional adaptive comfort model for Brazil in both the present and the future. Build Simul 17:459–482. https://doi.org/10.1007/s12273-023-1084-3
11. Sánchez-García D, Rubio-Bellido C, Tristancho M, Marrero M (2020) A comparative study on energy demand through the adaptive thermal comfort approach considering climate change in office buildings of Spain. Build Simul 13:51–63. https://doi.org/10.1007/s12273-019-0560-2
12. Escandón R, Sendra JJ, Suárez R (2015) Energy and climate simulation in the Upper Lawn Pavilion, an experimental laboratory in the architecture of the Smithsons. Build Simul 8:99–109. https://doi.org/10.1007/s12273-014-0197-0
13. Rubio-Bellido C, Pérez-Fargallo A, Pulido-Arcas JA, Trebilcock M (2017) Application of adaptive comfort behaviors in Chilean social housing standards under the influence of climate change. Build Simul 10:933–947. https://doi.org/10.1007/s12273-017-0385-9

14. Cuerda E, Guerra-Santin O, Sendra JJ, Neila González FJ (2019) Comparing the impact of presence patterns on energy demand in residential buildings using measured data and simulation models. Build Simul 12:985–998. https://doi.org/10.1007/s12273-019-0539-z
15. Aparicio Ruiz P, Barbadilla Martín E, Guadix Martín J, Muñuzuri J (2024) Predicting the clothing insulation through machine learning algorithms: a comparative analysis and a practical approach
16. Romero Rodríguez L, Sánchez Ramos J, Álvarez Domínguez S (2023) Building geometry data from online maps for accurate thermal simulations of districts
17. Ambiente M de M (2004) Water in Spain
18. Dolz J (2007) Problemática General del Drenaje de Aguas Pluviales en Zonas Urbanas. Barcelona: sn
19. Woods-Ballard B, Kellagher R, Martin P et al (2007) The SUDS manual. CIRIA, London
20. Gies E (2019) Ciudades esponja. In: Investigación y Ciencia
21. S.A.I.H. (2022) Confederación Hidrográfica del Guadalquivir. https://www.chguadalquivir.es/saih/. Accessed 3 June 2024
22. AEMET (2024) Base de datos Meteorológica. https://datosclima.es/. Accessed 3 June 2024
23. EMASESA (2019) INSTRUCCIONES TÉCNICAS PARA REDES DE SANEAMIENTO (PD 005.12) REVISIÓN 7:1–2
24. Balacco G, Carrieri AA (1967) Stima dei volumi di acque di prima pioggia e modalità di accumulo
25. UNI (2012) UNI_TS 11445_2012 - UNI Ente Italiano di Normazione.pdf
26. Palla A, Gnecco I, Lanza LG (2012) Compared performance of a conceptual and a mechanistic hydrologic models of a green roof. Hydrol Process 26:73–84. https://doi.org/10.1002/hyp.8112

Solar Photovoltaic Integration in Mediterranean Residential Buildings: Towards Sustainable Energy Solutions

Marta Torres González, Yanet Corona Macías, and Carlos Rubio Bellido

Abstract This chapter aims to evaluate the feasibility of supplying a building in Seville with photovoltaic (PV) energy to meet its heating, ventilation, and air conditioning (HVAC) and domestic hot water (DHW) needs according to the standards set by the Spanish Technical Building Code (CTE). Using the PVGIS calculator, it was found that complete energy supply could be achieved from December to March, with excess energy generated, but PV production falls short in the other months. Additionally, the study assesses HVAC and DHW energy consumption under various climate scenarios (RCP2.6, RCP4.5, and RCP8.5) for the years 2050 and 2100, revealing a slight decrease in heating demand and a significant increase in cooling demand due to climate change. The chapter also explores the potential of using higher efficiency PV panels available in the market to bridge the gap identified, comparing these findings with the current CTE specifications. This analysis highlights the need to reformulate the CTE to promote the construction of nearly zero-energy buildings and examines the implications of climate change on future energy consumption patterns.

1 Introduction

Since its establishment in 1988, the Intergovernmental Panel on Climate Change (IPCC) has published five comprehensive assessment reports, each focusing on various aspects of climate change, the increase in greenhouse gas (GHG) emissions, and the depletion of natural resources [1]. These reports provide critical insights into the evolving state of our climate and highlight the urgency of addressing environmental challenges [2–4].

The global decarbonization movement has driven the integration of renewable energy sources into existing power networks, advocating for a cleaner and more

M. Torres González (✉) · C. Rubio Bellido
Department of Architectural Constructions II, Universidad de Sevilla, Sevilla, Spain
e-mail: mtorres18@us.es

Y. Corona Macías
Universidad de Sevilla, Sevilla, Spain

M. Torres González and C. Rubio Bellido (eds.), *Towards Low and Positive Energy Buildings*, Green Energy and Technology,
https://doi.org/10.1007/978-3-031-70851-0_10

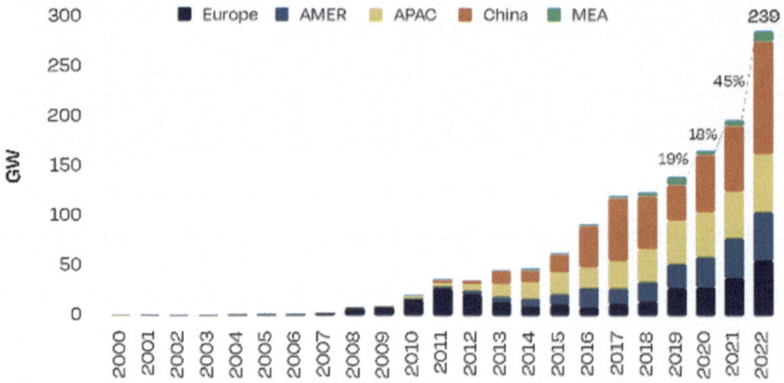

Fig. 1 Annual solar PV installed capacity 2000–2022 in the world [6]

centralized energy model. Over recent decades, technological advancements in renewable energy sources (RESs) have led to steady growth and significant penetration levels within the power grid [5]. The annual installed capacity of solar PV worldwide has increased notably from 2000 to 2022, as depicted in Fig. 1. This growth has reached 50 GW per year in Europe and 239 GW per year globally. The widespread use of these facilities to generate energy has continued to expand, demonstrating their viability and establishing them as a critical component of the global energy landscape.

However, this integration presents major challenges, particularly in managing and balancing power demand. Renewable technologies, characterized by their variable output, require innovative solutions to maintain the reliability and stability of power systems as they become more prevalent [5].

One notable opportunity arising from local generation is self-consumption, which enables consumers to use the energy they produce and potentially sell any surplus back to the grid. European legislation, specifically the Renewable Energy Directive (RED II) as part of the Clean Energy for All Europeans package (CEP), supports collective self-consumption. This approach typically involves individual energy producers who manage, operate, and distribute their production collectively. Photovoltaic (PV) technology is the most common method for generating renewable energy in collective local production set-ups or small distributed units, usually installed on rooftops or near the required power locations [5]. Self-consumption installations allow end users to prioritize covering their energy demands and sell any excess energy based on their load demand profiles, adhering to the legal frameworks and conditions of their respective countries [5]. For instance, installing PV systems on the rooftops of residential buildings for collective self-consumption transforms these rooftops into energy producers, significantly impacting energy demand through effective control of solar gains [7].

The importance of sustainable energy use in buildings cannot be overstated. Buildings are significant consumers of energy, and improving their energy efficiency is crucial for reducing overall energy consumption and GHG emissions. Regulations like the Spanish Technical Code (CTE) play a pivotal role in promoting energy efficiency in the building sector. The CTE sets out mandatory requirements for energy performance in new and refurbished buildings in Spain, aiming to enhance energy efficiency, integrate renewable energy sources, and ensure sustainable building practices. By adhering to these regulations, buildings can achieve lower energy consumption, reduced environmental impact, and increased comfort for occupants.

According to data retrieved from Solargis [8], an open-source platform, the potential for photovoltaic (PV) power generation varies significantly across the globe due to differences in solar resources. The global horizontal irradiation (GHI) is a key indicator of solar energy potential, representing the total amount of shortwave radiation received from the sun per unit area on a horizontal surface. Regions such as the Middle East, Northern Africa, and parts of Australia exhibit some of the highest GHI values, often exceeding 2200 kWh/m^2 per year (Figs. 2 and 3). These areas offer substantial opportunities for PV power generation due to their abundant and consistent solar resources.

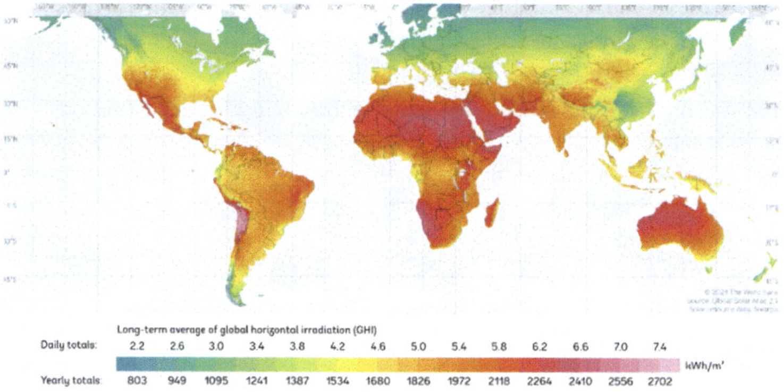

Fig. 2 Global horizontal irradiation. Retrieved from [8]

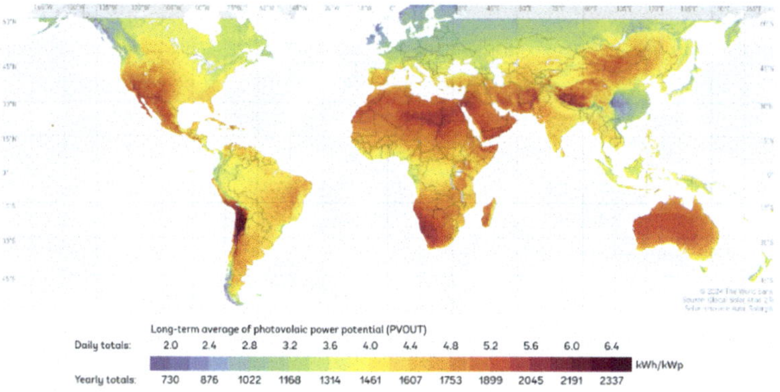

Fig. 3 Photovoltaic power potential. Retrieved from [8]

Recent studies highlight additional factors influencing PV performance. Tian et al. have shown that in countries with hot climates, a 1 °C rise in temperature results in a 1.66% increase in electricity consumption [9]. Furthermore, climate change projections indicate varying impacts on PV power production across Europe. While Scandinavian areas may experience a reduction of up to 10–12% due to climate change effects, Southern Europe, characterized by higher solar irradiance, is expected to maintain more favourable conditions for PV efficiency despite temperature effects on PV cell performance [10]. Conversely, regions with lower GHI values, such as parts of Northern Europe, still hold significant potential for PV deployment due to advancements in PV technology and efficiency. The Mediterranean area, particularly the south of Spain, also stands out with remarkably high GHI values (~ 5 kWh/m^2 per day), frequently surpassing 1800 kWh/m^2 per year, rendering it conducive for solar energy projects.

Southern Spain, with its high solar irradiance, presents an ideal environment for large-scale PV installations, contributing significantly to the country's renewable energy capacity and supporting Europe's broader goals for sustainable energy transition. The widespread availability of reliable solar radiation data from SOLARGIS underscores the feasibility and attractiveness of harnessing solar energy as a pivotal component of the global energy transition.

Recent studies highlight additional factors influencing PV performance. Tian et al. have shown that in countries with hot climates, a 1 °C rise in temperature results in a 1.66% increase in electricity consumption [9]. Furthermore, climate change projections indicate varying impacts on PV power production across Europe. While Scandinavian areas may experience a reduction of up to 10–12% due to climate change effects, Southern Europe, characterized by higher solar irradiance, is expected to maintain more favourable conditions for PV efficiency despite temperature effects on PV cell performance [10].

The International Energy Agency highlighted that heating and air conditioning (HVAC) systems, along with domestic hot water (DHW) systems, collectively account for nearly half of the global energy consumption in residential buildings [11]. In Seville, Andalusian capital, energy poverty looms large, affecting a significant segment of the populace [12]. High energy costs, coupled with low incomes and substandard housing, exacerbate this plight. Many families are compelled to choose between excessive energy expenditure and compromised thermal comfort. Limited access to efficient heating and cooling systems, alongside economic constraints inhibiting energy efficiency improvements, compound energy poverty in Seville [13]. Urgent interventions, encompassing enhanced energy efficiency, financial aid to low-income households, and equitable energy service access, are imperative to address this pressing issue comprehensively.

The objectives of this chapter are to evaluate the feasibility of using PV systems to meet the heating and air conditioning (HVAC) and DHW energy needs of a building in Seville according to the CTE guidelines. It aims to analyse the impact of various climate scenarios (RCP2.6, RCP4.5, and RCP8.5) on future energy consumption patterns for the years 2050 and 2100. Furthermore, the chapter investigates the potential advantages of employing higher efficiency PV panels relative to existing standards. This assessment aims to ascertain whether the building can achieve an energy consumption of less than 15 kWh/m^2 per year, a benchmark adopted by EU member states in the national transposition of the directive, thereby qualifying it as a nearly zero-energy building (NZEB) [14].

2 Case Study

The case study focuses on Andalusia, a large region in southern Spain with an approximate area of 88,000 km^2. This region encompasses several cities representative of the Mediterranean climate zone, characterized by hot, dry summers, and mild winters, including Seville. According to the Köppen climate classification, Andalusia fall within the Csa climate zone [15]. Specifically, the case study is located on the city of Seville, characterized by a central area covering 75,000 km^2, a population of 681,998, and an average per capita income of 13,085 in the year 2021 [16].

A specific residential high-rise building with flat roof was selected due to its construction typology, which has been commonly used in the region since the 1980s, in terms of dwelling type, compactness, dwelling area, and external envelope (Fig. 4) [17]. In this case, building analysed is taken from those constructed after the entry into force of the technical code in Spain during 2020 or from pre-existing buildings that have undergone rehabilitation in accordance with the envelope parameters established by the current Spanish technical code. The building was modelled using Design Builder software version 7.0.2.006. It features 11 floors, each one with 6 dwellings ranging from 100 to 115 m^2, designed for families of four members. The total building area spans up to 8121.86 m^2.

The thermal characteristics of the building envelope are regulated by the Spanish Technical Code [18]. For this purpose, maximum values of thermal transmittance (U value) were established for different envelope elements based on the winter climate severity (Table 1).

The rooftop area, measuring 659.80 usable square meters, is available and suitable for the installation of a photovoltaic (PV) system. The arrangement of PV solar modules in the case study was strategically planned to maximize coverage over the largest roof surface area, aligned parallel to the building walls (i.e. 390 m^2). Thus, the solar panels, which are composed of crystalline silicon cells, are positioned with a 37° tilt and an azimuth of 0° in a free-standing configuration, accounting for an 11% system loss. For this study, it is assumed that the building casts no shadows. However, it is also noted that there is an estimated 14% loss in energy production due to system inefficiencies and environmental factors.

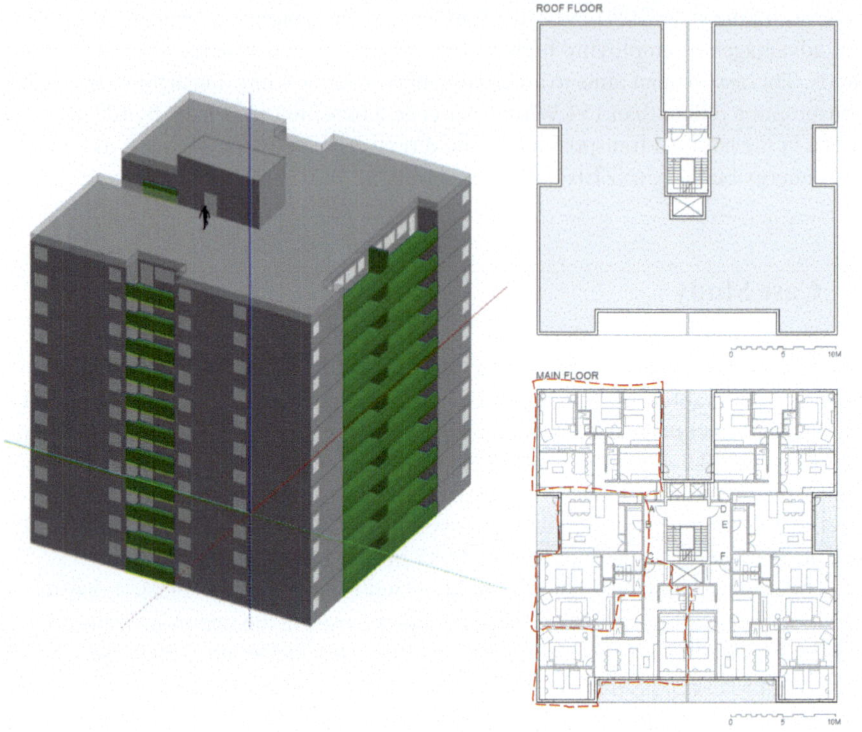

Fig. 4 Building modelled in ®DesignBuilder and plans of the studied block

Table 1 Limit *U*-values [W/(m^{2}k)] of building envelope since 2020 according to different winter climate severity [18]

Element	*U*-values [W/(m^{2}k)]
Wall	0.56
Elements in contact with the ground	0.75
Party wall	0.75
Roof	0.44
Floor in contact with the air	0.56
Window	2.30

3 Method

The EnergyPlus Weather files for Seville (referred to as EPW) were generated using METEONORM v8.1, and simulations were conducted using EP-Launch software, an optional component of EnergyPlus v9.1. These tools facilitated the simulation of the case study under current climatic conditions and future scenarios projected for 2050 and 2100, following IPCC guidelines (specifically, the RCP2.6, RCP4.5, and

RCP8.5 climate scenarios) [1]. The simulations provided data on cooling and heating energy demands as well as global horizontal irradiance (GHI) on a monthly basis.

To calculate the estimated refrigeration consumption, the guidelines specified in the Technical Building Code (CTE) have been followed (Eq. 1):

$$\text{Consumption}\left(\text{kW h/m}^2\right) = \frac{\text{Energy Demand}}{\eta} \tag{1}$$

where η is the equipment performance that acquires a value of 2.60 in summer periods and 0.92 in winter periods. However, in Seville, gas is falling out of use, and the heat pump offers numerous benefits, so coefficients corresponding to η:2.60 in summer (criterion taken from the CTE) and η:2.80 in winter (according to various commercial manufacturers, the performance of the equipment when operating in heating mode is + 0.2) will be adopted. The demand information was retrieved from the simulation using Ep-Launch for cooling and heating demand (kW h/m^2).

The estimated monthly DHW consumption for the entire building was 0.10 kWh/m^2, considering the water from the network starts at 10 °C and is subsequently heated to 60 °C for each dwelling.

The feasibility of installing a PV solar system on the building's roof and establishing an Energy Community was evaluated as part of the case study. This included sizing systems for electricity generation from renewable sources, specifically PV, for self-consumption and grid supply. The minimum installed inverter power, P_{min}, was determined as the lesser of two expressions (Eqs. 2 and 3), considering factors such as electrical production efficiency and roof area coverage by PV panels.

$$P_1(kW) = F_{\text{pr;el}} \cdot S \tag{2}$$

$$P_2(\text{kW}) = 0.1 \cdot (0.5 \cdot S_c - S_{\text{oc}}) \tag{3}$$

where $F_{\text{pr;el}}$ is the electrical production factor, which is valued at 0.005 for private residential use; S is the constructed area of the building (m^2); S_c is the roof area, and S_{oc} is the area occupied by PV panels (m^2). In this work, both P_{min} will be taken into consideration, to verify the effectiveness of the proposal in covering the installed peak PV power.

Additionally, to qualify as a collective self-consumption system with surplus energy and a simplified compensation model in accordance with national regulations, the maximum installed power must not exceed 100 kWp [19].

Once the consumption and the minimum and maximum power installed have been stated, the Photovoltaic Geographical Information System (PVGIS) [20] developed by the International Electrotechnical Commission (IEC) was used to pre-size mechanism for estimating the energy output of PV solar systems, incorporating various parameters to assess the case study. These parameters included: (i) Meteorological location data sourced from the SARAH2 satellite-based database, (ii) Determination of the horizon line in Seville, and (iii) PV system installation considerations such

Table 2 Characteristics of the main PV panels in Spain

Model	Efficiency (%)	Panel power (W)	Annual loss (%)	PV panel dimensions (mm)
LG NeON R	22.00	350–400	2.00	1700 × 1016
SunPower Maxeon 3	22.60	400	8.00	1690 × 1049
Panasonic HIT KURO	21.60	330–350	3.20	1580 × 798
REC Alpha Series	21.70	380	0.25	1755 × 1046
JinkoSolar Cheetah HC 72M-V	20.78	400–425	2.00	2066 × 1029

Bold indicates the maximum values

as roof surface area, tilt angle, azimuth of solar modules, and spacing between PV panel arrays to mitigate shadowing effects.

To conduct a comparative analysis of the CTE guidelines versus higher efficiency PV systems, commercial PV panels with enhanced technical specifications (Table 2) compared to those stipulated by the CTE were evaluated. This comparison aimed to identify how higher efficiency PV systems could bridge any performance gaps identified in the study, potentially enhancing energy generation and system reliability.

For this study, the monthly energy output of 172 commercial PV panels installed on the roof was analysed. Their power is 400 W, and the peak power is 69 kWp (i.e. $172 * 400 = 688{,}800$ Wp $\approx$ 69 kWp), considering a loss percentage of 2% and a panel efficiency of η:5 in summer and η:5.2 in winter (i.e. η:5 + 0.20) (Fig. 8).

4 Results and Discussion

The energy performance of the case study was analysed with limit values established in the CTE. The spatial distribution of the values of heating and cooling energy consumption in the case of study were depicted in Fig. 5. These figures reveal that high energy consumption is primarily associated with cooling during the autumn, spring, and summer months. Thus, the months of July (7.655 kWh/m^2 ± 0.96 SD), August (7.354 kWh/m^2 ± 0.93 SD), and September (6.058 kWh/m^2 ± 0.98 SD) exhibit the highest average energy consumption across the various climate scenarios studied.

As winter and summer climates became more severe, there was a notable upward trend in energy demand. Specifically, the demand for cooling substantially increased in the present, 2050, and 2100 scenarios. Conversely, the demand for heating energy (for DHW and heating) slightly decreased overall, with only slight increases during peak winter months in December and January (Table 3).

Fluctuations in energy consumption resulted in variations ranging from – 0.47004 to – 9.3764 kWh/m^2 per year for cooling and from 0.0912 to 0.4661 kWh/m^2 per

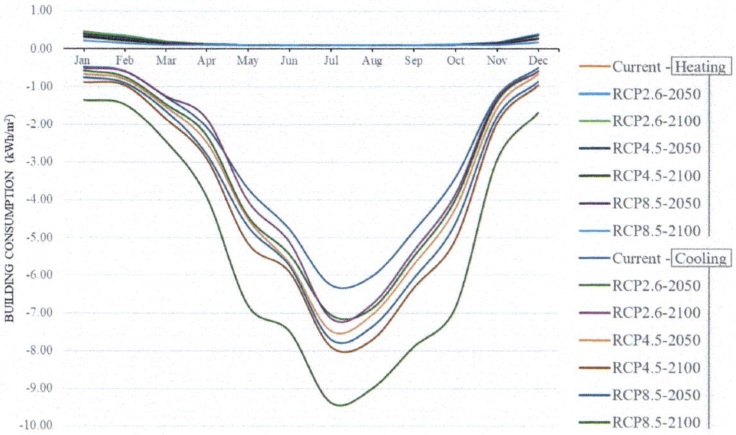

Fig. 5 Monthly building consumption in heating (DWH + air conditioning) and cooling

Table 3 Annual energy consumption of the case study (kWh/m^{2}year)

Scenarios	Cooling consumption				Heating consumption			
	Min.	Max.	Avg.	SD	Min.	Max.	Avg.	SD
Current	– 0.470	– 6.258	– 2.928	2.182	0.091	0.466	0.191	0.131
RCP2.6-2050	– 0.581	– 7.047	– 3.364	2.475	0.091	0.438	0.183	0.124
RCP4.5-2050	– 0.662	– 7.468	– 3.526	2.550	0.091	0.396	0.174	0.110
RCP8.5-2050	– 0.759	– 7.687	– 3.745	2.595	0.091	0.345	0.157	0.087
RCP2.6-2100	– 0.506	– 7.137	– 3.193	2.480	0.091	0.447	0.188	0.125
RCP4.5-2100	– 0.882	– 7.881	– 3.959	2.671	0.091	0.313	0.149	0.076
RCP8.5-2100	– 1.358	– 9.376	– 5.103	3.089	0.091	0.215	0.121	0.037

year for heating. This observation indicates that certain scenarios, such as RCP 8.5, exhibit traits conducive to higher energy demands.

As reflected in the previous studies, heating consumption slightly decreases but does not constitute a significant change, whereas the building's cooling consumption tends to increase significantly in future climate scenarios. This trend indicates that any analysis we conduct in the current state must consider these projections (Fig. 6).

4.1 Balance Between Energy Consumption and Energy Production in the Building

Considering that the installed PV system covers the minimum and maximum peak power (100 kWp), with 11% losses and a 35° tilt, which is deemed optimal by GIS for

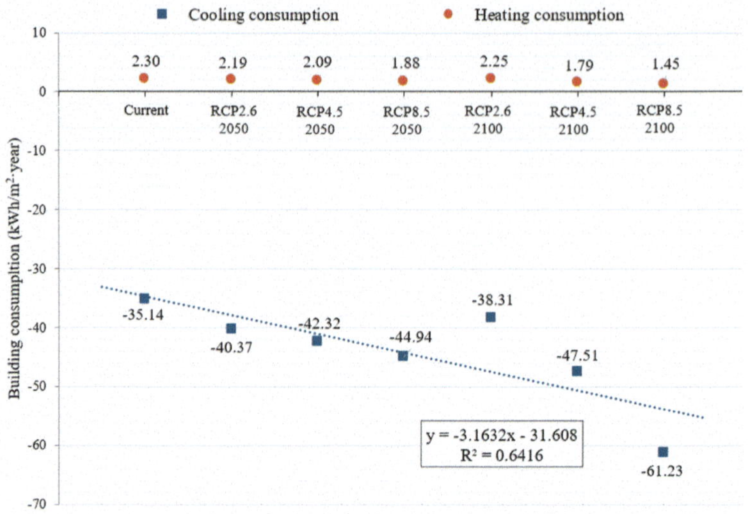

Fig. 6 Building consumption aligns with the performance criteria specified in the CTE

the city of Seville, the PV solar installation sizing and simulation of the building was performed using the PVsyst software package. The results obtained indicate that if the PV installation meets the minimum power requirements according to CTE, it covers 25% of the building's annual consumption. In contrast, if the PV installation meets the maximum power requirements according to CTE, it covers 56% of the building's annual consumption for DHW, heating, and cooling. Additionally, it generates an energy surplus in December, January, February, and March, which can be stored and used in the summer months when cooling energy consumption peaks (Fig. 7).

The results shown in Fig. 7 were compared with the monthly energy output of 172 commercial 400 W panels installed on the roof of the case study. In this situation, the peak power is 69 kWp (i.e. $172 * 400 = 688{,}800$ Wp $\approx$ 69 kWp), considering a loss percentage of 2% and panel efficiency of η:5 in summer and η:5.2 in winter (i.e. η:5 + 0.20) (Fig. 8).

Initially, it is observed that maintaining the demand while nearly doubling the performance of the PV installation significantly reduces the building's monthly energy consumption for climate control and domestic hot water (DHW). Furthermore, the production curve of commercial PV panels closely aligns with the maximum production curve of a 100 kWp PV system as per the CTE criteria. Specifically, the new production curve is approximately 3000–4000 kWh less per month than the former (Fig. 8).

The PV panel system covers up to 81.74% of the annual energy consumption for cooling, heating, and DHW. It fully meets the consumption needs from November to April. During these months, the total consumption is 16,786.53 kWh, while the production reaches 26,904.36 kWh, resulting in an energy surplus of 10,117.83 kWh

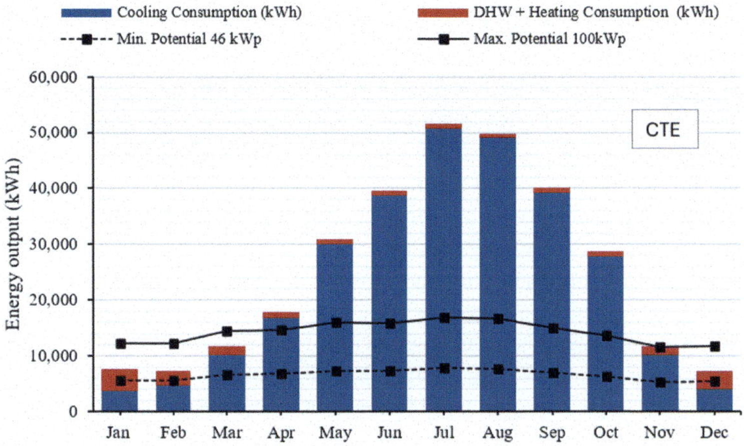

Fig. 7 PV panels according to CTE criteria. Monthly energy output from fix-angle PV system VS building consumption. Retrieved from [20]

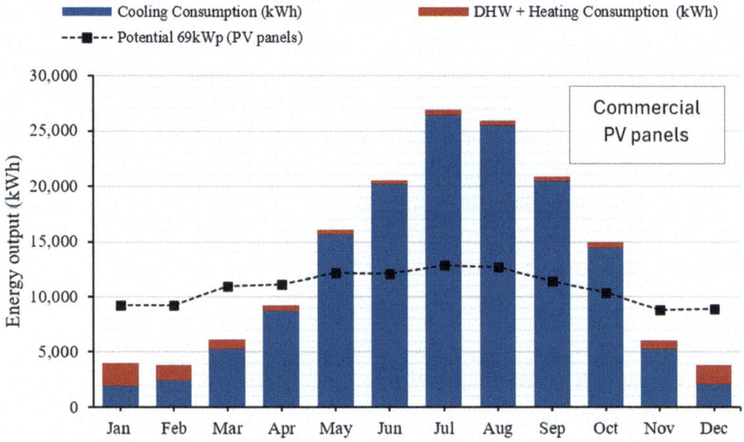

Fig. 8 Commercial PV panels. Monthly energy output from fix-angle PV system VS building consumption. Retrieved from [20]

(Fig. 8). This excess energy can be reused for building electricity and lighting or returned to the grid.

Based on the comparison between Figs. 7 and 8, the following conclusions can be drawn:

- The CTE is currently not reflecting the latest innovations and technological advancements in the field of PV energy, thereby limiting the use of the most efficient and cost-effective technologies available in the market.

- PV energy is unable to supply electricity continuously over time and is also dependent on the weather. This means that in the face of volatile energy demand with peaks, photovoltaic energy has to ensure the supply through storage or alternative energies. The CTE also has limited consideration for the integration of PV systems with other energy generation and storage technologies, directly restricting the implementation of hybrid solutions that could optimize energy consumption and enhance the resilience of buildings against fluctuations in energy demand and supply.
- The use of PV self-consumption systems allows users to reduce their electricity bill since self-production energy is not extracted from the grid and, therefore, is not billed. Consequently, the savings derived from the reduction of the electricity bill allow users to recover the installation costs within a period of 5–10 years, depending on solar radiation and the location of the system. However, the CTE lacks clear mechanisms to compensate for excess energy production, and a detailed hourly study of PV energy consumption would be necessary.

In this regard, it is worthwhile to mention that previous studies compare the economic profitability proposed by the current situation of collective photovoltaic self-consumption in Seville with a residential building without PV installation. The energy produced is used for self-consumption, and during months with excess production, it can be returned to the electrical grid, obtaining compensation through reduced bills or economic benefits by selling the excess generated directly to the utility company or through peer-to-peer sales, selling surplus energy directly to other local consumers, promoting a more decentralized and efficient energy market. This results in monthly economic savings on the household electricity bill, thereby favouring the mitigation of energy poverty and contributing to emissions reduction [21].

4.2 *Impact of Climate Conditions and Climate Change on PV Panel Performance*

The operation of PV panels is largely conditioned by the climatic conditions of the region where they are installed, so variations in these conditions due to climate change could modify the performance of the systems in future climatic scenarios. Failure to consider this reality by the CTE could lead to projections and designs that are inadequate and fail to optimize long-term energy production.

In Mediterranean regions like Southern Spain, while an increase in global solar radiation could mean a rise in the potential energy production of PV panels, during summer periods with extreme temperatures, the combination of higher solar radiation and very high ambient temperatures could lead to a significant increase in the surface temperature of the photovoltaic cell. This increase in temperature can raise circuit resistance and lower voltage, resulting in decreased system efficiency [22]. In this regard, the surface temperature of a module has been described in the literature through the expression (Eq. 4) [23]:

$$T_{\text{mod}}\,(^{\circ}\text{C}) = T_{\text{amb}} + \text{Irradiance} \cdot \exp\left(-a - b \cdot \text{WS}\right) + \Delta T \frac{\text{Irradiance}}{1000} \tag{4}$$

where T_{amb} is the ambient temperature (°C), WS is the wind speed (m/s), and a, b, and ΔT are constants that depend on the material of the solar cell.

Figures 9 and 10 show monthly average outdoor air temperature and average global horizontal radiation for the Seville region under current conditions and future scenarios (RCP 2.6, 4.5, and 8.5) for the years 2050 and 2100.

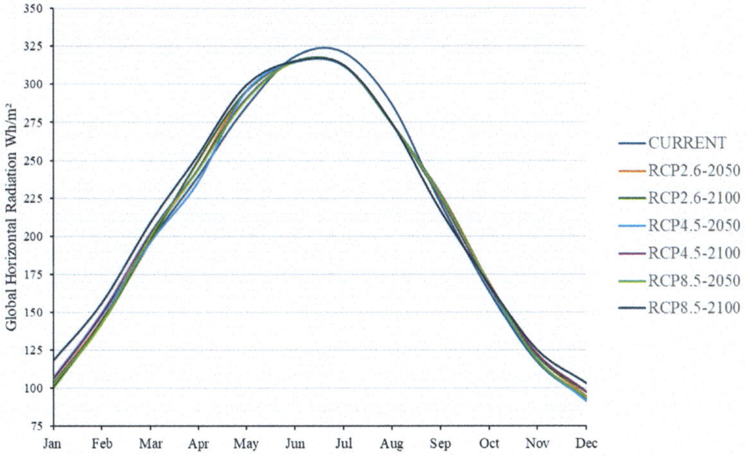

Fig. 9 Monthly average global horizontal irradiation (Wh/m^2) in Seville

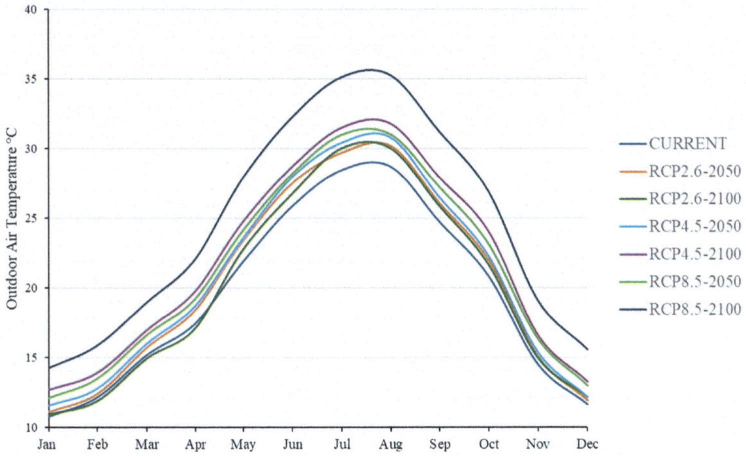

Fig. 10 Monthly average outside air temperature (°C) in Seville

Table 4 Outdoor air temperature, wind speed, and global horizontal radiation in July in Seville

		Current	2050			2100		
			RCP 2.6	RCP 4.5	RCP 8.5	RCP 2.6	RCP 4.5	RCP 8.5
Air temp. (°C)	Min.	16.40	17.50	18.20	19.00	17.70	19.40	22.60
	Max.	42.50	43.70	44.00	45.20	43.80	45.60	48.80
	Avg.	28.40	29.70	30.40	31.00	30.00	31.50	35.10
Wind speed (m/s)	Min.	0.00	0.00	0.00	0.00	0.00	0.00	0.00
	Max.	14.50	14.50	14.50	14.50	14.50	14.50	14.50
	Avg.	2.80	2.80	2.80	2.80	2.80	2.80	2.80
Global horizontal radiation (Wh/m^2)	Min.	0.00	0.00	0.00	0.00	0.00	0.00	0.00
	Max.	1,039.00	1,039.00	1,038.00	1049.00	1021.00	1,027.00	1,025.00
	Avg.	321.00	312.30	312.50	312.50	312.60	312.50	312.50

While it might be considered a positive aspect for PV energy production that in future scenarios, the greatest increase in solar radiation occurs during winter months while remaining nearly constant or lower in summer, the rise in outdoor air temperatures by approximately 25% by 2100 in the RCP 8.5 scenario for summer months, with an increase of nearly 7 °C in July compared to current temperatures, could result in decreased system performance.

Given that an increase in cell temperature above 36° [24] leads to reduced electrical efficiency when cooling systems are not in place, with literature suggesting a decline of 0.3–0.5% per °C increase [25], and considering the findings from our analysis of different scenarios, particularly under RCP 8.5 for 2100, where summer months show an average monthly outdoor air temperature reaching 35.08 °C in July, peaking at 48.80 °C (Table 4), while other factors like global horizontal radiation and wind speed remain relatively stable, it is evident that photovoltaic panel performance in the region will be impacted by climate change. This underscores the need for panels designed for efficient operation in high temperatures, potentially incorporating cooling systems.

These deficiencies underscore the necessity to continuously review and update the Spanish standard to incorporate best practices, emerging technologies, and policies that promote the effective and sustainable adoption of solar photovoltaic energy in buildings.

5 Main Conclusions and Future Directions of Research

This study examined the energy performance of a residential case study within the framework of the Spanish Technical Building Code (CTE), focusing on heating, cooling, and domestic hot water (DHW) consumption across different climate scenarios (current, 2050, and 2100) based on IPCC guidelines. The results revealed a notable rise in cooling demand across all future scenarios, particularly during severe summer conditions, while heating demand either decreased slightly or remained stable. Energy consumption displayed significant variability depending on the climate scenario, with scenarios like RCP 8.5 showing higher energy demands, especially for cooling.

The analysis underscored that the primary driver of high energy consumption in the case study was the need for cooling during July, August, and September, highlighting the seasonal fluctuation in energy demand and the critical influence of climate conditions on consumption patterns. Evaluating the efficacy of the PV system in meeting energy needs, the study found that a properly sized PV installation under CTE criteria could cover between 25 and 56% of the building's annual energy consumption. This emphasizes the potential for renewable energy systems to substantially reduce building energy dependence, particularly during months of surplus energy production.

The findings suggest that current building codes, while sufficient, may require updates to accommodate future climate conditions and advancements in renewable energy technologies. There exists an opportunity to optimize energy performance through revised regulations that integrate more efficient and resilient building practices, including enhanced PV systems and energy storage solutions.

Drawing from this study, achieving Nearly Zero Energy Buildings (NZEB) necessitates leveraging renewable energy sources like PV systems and updating regulatory frameworks to align with forthcoming climate challenges and technological innovations. Comprehensive assessments of building energy that consider both current and projected climate impacts can promote sustainable building practices and diminish environmental footprints.

While this study provides clarity on residential PV energy consumption, further research is necessary to analyse costs and impacts on average household energy expenditures, addressing energy poverty and advocating for solar adoption based on resident income levels. It is crucial to account for a mismatch factor related to hidden energy poverty, where vulnerable users reduce energy consumption to lower housing costs, resulting in discomfort from extreme temperatures and limited use of hot water.

Climate change will impact the efficiency, durability, and performance of photovoltaic panel systems in Spain. Technological adaptation and mitigation strategies will be essential in addressing these challenges and optimizing solar energy production in the future. The long-term impact will also hinge on technological advancements in photovoltaic panel manufacturing. New technologies have the potential to be more resilient to high temperatures and changes in solar radiation, thereby mitigating some adverse effects of climate change.

References

1. IPCC (2014) Climate change 2014: synthesis report. Contribution of working groups I, II and III to the fifth assessment report of the intergovernmental panel on climate change

2. Kalvelage K, Passe U, Rabideau S, Takle ES (2014) Changing climate: the effects on energy demand and human comfort. Energy Build 76:373–380. https://doi.org/10.1016/j.enbuild.2014.03.009
3. Wang H, Chen Q (2014) Impact of climate change heating and cooling energy use in buildings in the United States. Energy Build 82:428–436. https://doi.org/10.1016/j.enbuild.2014.07.034
4. Rubio-Bellido C, Pérez-Fargallo A, Pulido-Arcas JA (2016) Optimization of annual energy demand in office buildings under the influence of climate change in Chile. Energy 114:569–585. https://doi.org/10.1016/j.energy.2016.08.021
5. Parreño-Rodriguez A, Ramallo-González AP, Chinchilla-Sánchez M, Molina-García A (2023) Community energy solutions for addressing energy poverty: a local case study in Spain. Energy Build 296.https://doi.org/10.1016/j.enbuild.2023.113418
6. Global market outlook for solar power 2023–2027 (2024)
7. Bienvenido-Huertas D, Sánchez-García D, Tejedor B, Rubio-Bellido C (2023) An innovative approach to assess the limitations of characterizing solar gains in buildings: a Spanish case study. Energy Build 293.https://doi.org/10.1016/j.enbuild.2023.113206
8. SOLARGIS (2024) The World Bank, solar resource data: Solargis. https://solargis.com/es. Accessed 10 June 2024
9. Tian L, Li Y, Lu J, Wang J (2021) Review on urban heat island in china: methods, its impact on buildings energy demand and mitigation strategies. Sustainability 13:1–31. https://doi.org/10.3390/su13020762
10. Jerez S, Tobin I, Vautard R, et al (2015) The impact of climate change on photovoltaic power generation in Europe. Nat Commun 6.https://doi.org/10.1038/ncomms10014
11. IEA International Energy Agency. Heating. https://www.iea.org/energy-system/buildings/heating. Accessed 10 Apr 2024
12. Bienvenido-Huertas D, Sanz Fernández A, Sánchez-Guevara Sánchez C, Rubio-Bellido C (2022) Assessment of energy poverty in Andalusian municipalities. Application of a combined indicator to detect priorities. Energy Rep 8:5100–5116. https://doi.org/10.1016/j.egyr.2022.03.045
13. Alba-Rodríguez MD, Rubio-Bellido C, Tristancho-Carvajal M et al (2021) Present and future energy poverty, a holistic approach: a case study in Seville, Spain. Sustainability 13:1–15. https://doi.org/10.3390/su13147866
14. Parliament E (2010) Directive 2010/31/EU of the European Parliament and of the Council of 19 May 2010 on the energy performance of buildings (recast)
15. Arnfield AJ (2023) Köppen climate classification. In: Encyclopedia Britannica. https://www.britannica.com/science/Koppen-climate-classification. Accessed 17 July 2023
16. Instituto Nacional de Estadística (2024). https://www.ine.es/. Accessed 6 June 2024
17. Ordóñez J, Jadraque E, Alegre J, Martínez G (2010) Analysis of the photovoltaic solar energy capacity of residential rooftops in Andalusia (Spain). Renew Sustain Energy Rev 14:2122–2130. https://doi.org/10.1016/j.rser.2010.01.001
18. Ministerio de Transportes Movilidad y Agenda Urbana (2022) Documento Básico HE. Ahorro de Energía
19. Ministerio para la Transición Ecológica (2019) Real Decreto 244/2019 de 5 de Abril 2019
20. European-Commision (2022) Photovoltaic geographical information system (PVGIS). https://re.jrc.ec.europa.eu/pvg_tools/en/. Accessed 17 May 2024
21. Jiménez Spínola, G (2020) Análisis de escenarios de autoconsumo fotovoltaico en España. Universidad de Sevilla https://biblus.us.es/bibing/proyectos/abreproy/93271. Accessed 16 September 2024
22. Sharaf M, Yousef MS, Huzayyin AS (2022) Review of cooling techniques used to enhance the efficiency of photovoltaic power systems. Environ Sci Pollut Res 29:26131–26159. https://doi.org/10.1007/s11356-022-18719-9
23. Hasan K, Yousuf SB, Tushar MSHK et al (2022) Effects of different environmental and operational factors on the PV performance: a comprehensive review. Energy Sci Eng 10:656–675. https://doi.org/10.1002/ese3.1043

24. Katkar AA, Shinde NN, Patil PS (2011) Performance & evaluation of industrial solar cell w.r.t. temperature and humidity. Int J Res Mech Eng Technol 1:69–73
25. Nižetić S, Čoko D, Yadav A, Grubišić-Čabo F (2016) Water spray cooling technique applied on a photovoltaic panel: the performance response. Energy Convers Manage 108:287–296. https://doi.org/10.1016/j.enconman.2015.10.079

Assessing Heat Island Growth in a Coastal City on the Yucatan Peninsula Using Geographic Information System

Mario Jiménez Torres, Román A. Canul Turriza, Oscar May Tzuc, Andrea Cruz y Cruz, Sandra J. Figueroa Ramírez, and Roselia Turriza Mena

Abstract This chapter explores the application of geographic information systems (GIS) to analyse and project urban heat islands (UHIs) in the Yucatan Peninsula. Currently, UHIs are a significant problem in developing cities, especially in regions with tropical climates. The chapter presents a detailed methodology to detect UHI's presence and temporal growth from the processing and treatment of satellite images supported by climatological data measured in situ. Take the City of Francisco de Campeche case study, which has witnessed remarkable urban growth in the last decade and is estimated to continue in subsequent years. The surface temperature of the city and the vegetation index are analysed, and it is observed that it is possible to correlate both phenomena and indicate the increase in temperature associated with it. For the case study, the increase was 1.4 °C. Finally, the chapter presents proposals for mitigating the increase in temperature in the peninsular city based on reforestation and how these are linked to previous studies.

1 Introduction

Climate change and global warming are critical concerns for society product of the significant threat to the planet's ecosystems and quality of life; for that reason, the reduction of pollutants and environmental degradation is part of the objectives of each government [1]. The Department of Environmental Information of the National

M. Jiménez Torres · R. A. Canul Turriza (✉) · O. May Tzuc · A. Cruz y Cruz · R. Turriza Mena
Facultad de Ingeniería, Universidad Autónoma de Campeche, Campus V, S/N Por Av. Humberto Lanz Cárdenas y Unidad Habitacional Ecológica Ambiental, Col. Ex Hacienda Kalá, 24085 San Francisco de Campeche, Campeche, México
e-mail: roacanul@uacam.mx

S. J. Figueroa Ramírez
Facultad de Ingeniería, Universidad Autónoma del Carmen, Av. Concordia, Carmen Campeche, 24180 Ciudad del Carmen, Campeche, México

M. Torres González and C. Rubio Bellido (eds.), *Towards Low and Positive Energy Buildings*, Green Energy and Technology,
https://doi.org/10.1007/978-3-031-70851-0_11

Office of Oceanic and Atmospheric Administration (NOAA) revealed in the Global Climate Report that the monthly temperature in every month during 2021 exceeded the average temperature recorded in previous years [2]. Urban heat islands (UHIs) have reinforced this phenomenon in recent decades [3]. This thermal anomaly significantly increases the temperature in urban areas, directly impacting the population and the environment [4]. Thus, understanding the UHI is crucial for mitigating its effects and ensuring the well-being of urban dwellers [5].

Multiple investigations have established a direct influence between UHI and air pollution, aggravating its behaviour in urban regions. This correlation is due to elevated surface temperatures and pollutants from dense vehicular traffic, which produce particulate matter, nitrogen oxides, and ozone [6]. Therefore, the adverse effects on health in cities are considerable. According to the World Health Organization (WHO), air pollution causes 4.2 million deaths annually, while indoor air pollution causes approximately 3.8 million deaths [7]. Likewise, it is estimated that 91% of the population lives where the air quality index exceeds the WHO limits; therefore, regulating urbanization can present two-way benefits [6, 7].

Urbanization is accompanied by significant environmental changes, including vegetation, soil, and climate [8]. The growth of the urban population, economic development, and changes in land use increase the effects of UHI, generating negative effects in multiple aspects [9]. This is relevant because cities currently represent 55% of the world's population, which is expected to increase to 68% by 2050 [10]. Therefore, it is essential to understand how the components of cities interfere with the UHI to establish improvement measures in the urban thermal environment and reduce air pollution [11]. Among these components, the fractal dimension of urban planning, which controls population density, has an important role. Optimizing urban geometry can decrease the effect of UHI; however, it can also present collateral benefits in the air and thermal environment. Therefore, focusing on urban decentralization, expansion control, green coverage rate, and building density will improve the thermal environment and air pollution [11, 12].

In coastal regions, the effect of the UHI is extreme, modifying the regional meteorology from extreme heat waves to floods, and the construction trend and human activities are expected to intensify the phenomenon [13]. Furthermore, proximity to the coast in these regions increases climatic complexity due to phenomena such as high climate variability, cyclones, and anticyclones, among others [14]. The above phenomenon is a product of the sea breeze, which directs the effects of UHI several kilometres inland towards the city until it dissipates [15]. Therefore, it is necessary to understand the phenomenology allowing the formulation of policies supporting decision-making and scenario planning, which consider (a) Analysis of the time scale; (b) Inclusion of the landscape and urban form, proportion of green and blue areas, improvement of albedo, and modal distribution of transportation; (c) passive technologies in the envelope of buildings and cold surfaces; (d) active technologies considering sustainable transportation under the criteria of energy consumption, heating, ventilation, refrigeration, and residual heat; and (e) public health and citizen participation [16].

Currently, there is limited knowledge about the spatiotemporal variation of surface UHI intensity during daytime and night-time. Additionally, more resources are needed to address the rapid impacts of urbanization. This study aims to identify and characterize the UHI in San Francisco de Campeche using satellite images, leveraging its coastal location and substantial growth in construction spaces to propose mitigation actions.

2 Case of Study

Mexico is located in the southeastern of North America and covers nearly two million square kilometres [17]. Due to its vast territory, varied latitude, and longitude, among others, the country exhibits significant climatic diversity. According to the Köppen–Geiger classification, Mexico has over 60 climate types [18]. Climatic diversities are distributed across the 32 states. In this context, San Francisco de Campeche, the capital of Campeche (19° 50′ 41” N, 90° 32′ 23” W), located in the peninsular region on the shores of the Gulf of Mexico, presents an interesting case for analysis (Fig. 1). This interest arises from its population increase and distribution, influenced by its designation as a UNESCO World Heritage site in 1999, recognized for its military, civil, and religious architecture from the sixteenth to nineteenth centuries. San Francisco de Campeche, known colloquially as 'The Bride of the Seas', has a territorial extension of 3410.64 km^2 (6% of the total state territory) with an average altitude of 5 m above the sea level (Fig. 2). The average annual temperature is 27 °C, with summer highs exceeding 30 °C and a historical maximum of 43 °C. The region experiences a warm-humid climate characterized by three seasons: the rainy season from June to the end of September, the 'Nortes' or winter storm season from October to January, and the dry season from February to May [19]. The average annual rainfall is 1096.7 mm, the highest from July to September [20].

The city has experienced significant growth, with its population increasing from 190,813 in 2000 to 294,077 in 2020, an increase of 54.11% [21]. This population surge has led to unplanned urbanization, primarily due to land invasions that were converted into residential areas, resulting in a considerable reduction of green spaces within the city. These areas are identified as potential UHIs, which have led to increased use of air conditioning, higher energy demand, and greater generation of atmospheric pollutants (CO_2, NO, SO_2, and PM). Therefore, UHIs are identified and characterized in the City of San Francisco de Campeche, and mitigation actions are proposed.

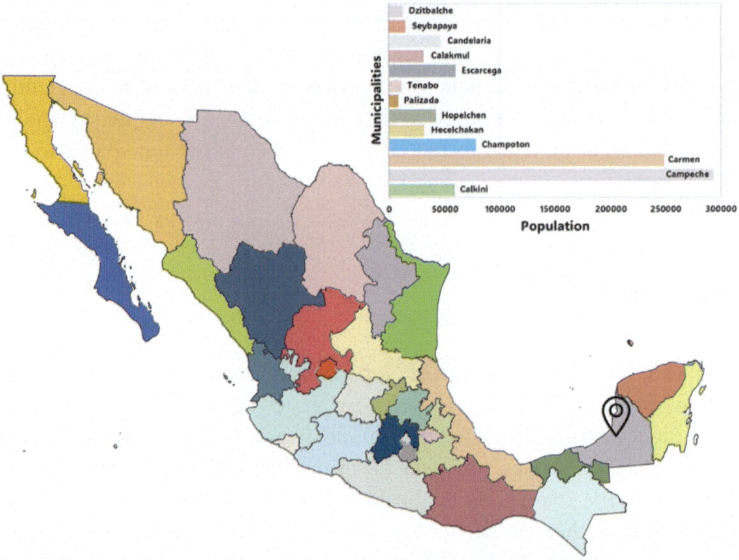

Fig. 1 Political division of the states of Mexico and population

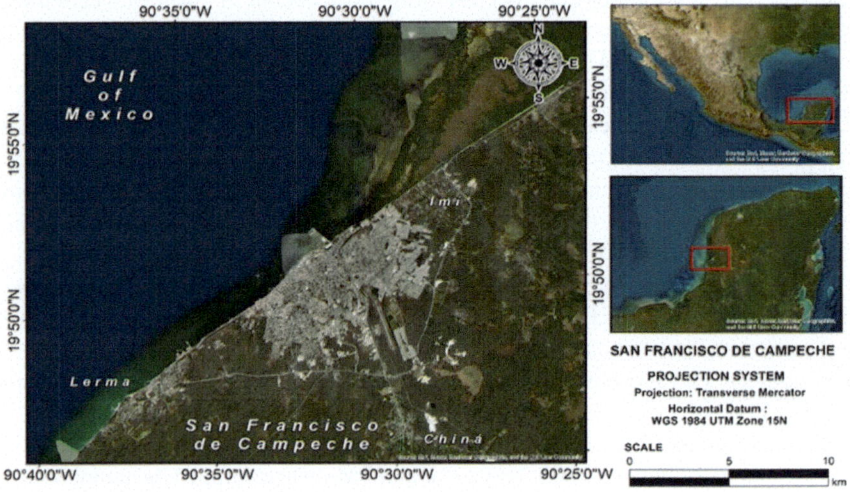

Fig. 2 Location of the State of Campeche

3 Methodology

Figure 3 illustrates the methodology used to analyse the city's UHI and the procedure for reforestation proposal. The analysis includes three temporal elements: land surface temperature (LST), normalized difference vegetation index (NDVI), and an examination of the local temperature history record. To calculate the LST and NDVI, Landsat satellite images were used, which are represented through different spectral bands with resolutions of 30 × 30 m. The study utilized data from Landsat-5TM, Landsat-7TM, Landsat-8OLI, and Landsat-9OLI obtained from the United States Geological Survey (USGS). The images were examined from 1990 to 2020 at 5-year intervals, focusing on April to characterize the dry season, which is the hottest in the region. Additionally, the historical climate analysis was conducted using data from the ERA5 model dataset created by the European Centre for Medium-Term Weather Prediction and local weather stations.

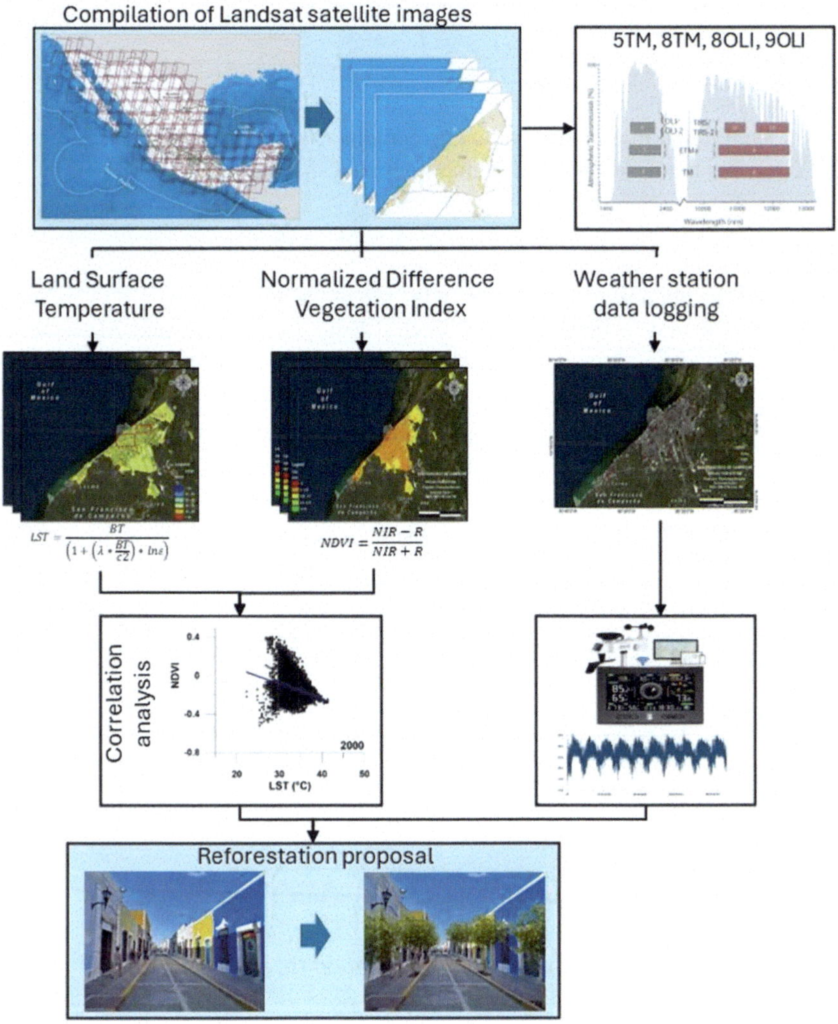

Fig. 3 Workflow of the proposed methodology

3.1 Land Surface Temperature

It is computed by using the images from band 6 (thermal infrared) for Landsat-5TM and Landsat-7TM, and spectral band 10 (infrared sensor) from Landsat-8OLI and Landsat-9OLI were used. The calculation for obtaining land surface temperature from satellite images consists of 5 steps summarized below [22]: Spectral radiance (L_λ) for TM images (Eq. 1), spectral radiance (L_λ) for OLI images (Eq. 2), brightness temperature (BT) (Eq. 3), land surface emissivity (LSE) (Eq. 4), and land surface temperature (LST) (Eq. 5):

$$L_\lambda = \left(\frac{L_{\max,\lambda} - L_{\min,\lambda}}{Q_{\text{cal},\max} - Q_{\text{cal},\min}}\right) * \left(Q_{\text{cal}} - Q_{\text{cal},\min}\right) + L_{\min,\lambda} \tag{1}$$

where Q_{cal} is the digital value of the pixel in an interval 0–255, $Q_{\text{cal},\max}$ and $Q_{\text{cal},\min}$ the maximum and minimum pixel value contained in the thermal band, and $L_{\max,\lambda}$ and $L_{\min,\lambda}$ are the maximum and minimum scaled spectral radiances.

$$L_\lambda = \text{ML} * Q_{\text{cal}} + \text{AL} - O_i \tag{2}$$

It considers obtaining the radiation at the top of the atmosphere or TOA radiance, where O_i is the correction value for spectral band 10, and ML and AL represent multiplicative and additive factors, respectively.

$$\text{BT} = \frac{K2}{\ln\left(\frac{K1}{L_\lambda} + 1\right)} - 273.15 \tag{3}$$

Expressed in °C, where $K1$ and $K2$ are known as the thermal conversion constant and are associated with the type of satellite image (TM or OLI).

$$\text{LSE} = 0.004 * \left(\frac{\text{NDVI} - \text{NDVI}_{\min}}{\text{NDVI}_{\max} - \text{NDVI}_{\min}}\right)^2 + 0.986 \tag{4}$$

Indicates the average emissivity of an element on the ground surface, where $\text{NDVI}_{\min}$ and $\text{NDVI}_{\max}$ are the maximum and minimum values of the normalized difference vegetation index (NDVI).

$$\text{LST} = \frac{\text{BT}}{\left(1 + \left(\lambda * \frac{\text{BT}}{h*c/s}\right) * \ln(\text{LSE})\right)} \tag{5}$$

where λ is the wavelength of the emitted radiation (in μm), h represents $6.626 \times 10^{(-34)}$ JS, s refers to the Boltzmann constant ($1.38 \times 10^{(-23)}$ J/K), and c is the speed of light (2.998×10^8 m/s).

3.2 Normalized Difference Vegetation Index

The normalized difference vegetation index (NDVI) is commonly used to check vegetation conditions by calculating the ratio between the Landsat imagery's near-infrared (NIR) and red (*R*) bands. The estimation of the indicator is given by Eq. 6 [22]:

$$\mathrm{NDVI} = (\mathrm{NIR} - R)/(\mathrm{NIR} + R) \tag{6}$$

where NIR corresponds to Band 4 (Landsat 5 and 7) and Band 5 (Landsat 8 and 9), while *R* corresponds to Band 3 (Landsat 5 and 7) and Band 4 (Landsat 8 and 9). NDVI values range from − 1 to + 1, depending on the objects visible at ground level; typically, green vegetation is between 0.2 and 0.8 [23].

3.3 Local Temperature History Record

Consists of a temporal record from 1940 to 2023 based on data compiled from two sources. On the one side, the data from 1940 to 2022 was obtained from the ERA5 model [24], the fifth-generation ECMWF reanalysis for the global climate and weather for the past eight decades. Data is available from 1940 onwards. ERA5 replaces the ERA-Interim reanalysis, data on the air temperature at 2 m above the earth's surface, to identify if the temperature increases over time in the city. This value is calculated by interpolating between the lowest level of the model and the Earth's surface, considering atmospheric conditions. On the other hand, weather data from 2022 to 2023 were obtained from a multifunctional wireless weather station located within the city of San Francisco de Campeche at coordinates 19.85° N–90.50° W that records data on climatological variables: temperature, humidity, barometric pressure, dew point, wind speed and direction, precipitation, solar, and UV radiation; the temperature data series from October 2022 to April 2023 was obtained with a record every 10 min.

Additionally, the previous information was complemented with The Municipal Urban Development Program of Campeche 2020–2040 (PMDU), which presents background information on the planning of urban growth in the city of San Francisco de Campeche as well as the existing territorial planning instruments: the Urban Development Master Program of Campeche 2008–2033 (PDU) and the Municipal Program of Ecological Planning of the Territory (PMOET). It also presents the growth of the population and its relationship with the urbanized area.

4 Results

4.1 Analysis of Historical Population Growth

Figure 4 illustrates the trends in demographic growth, the number of residential buildings constructed, and the impact on urban fragmentation in Campeche from 1950 to 2019. The figure reveals an 87% increase in population over seventy years, with the most significant spike occurring in the last thirty years, from 1970 to 2019. However, it is important to clarify that a third of this population increase has occurred in the last ten years. For its part, the growth linked to the number of homes required by the population has been greater, increasing by 91% from 1950 to date, highlighting that only since 2000 has the expansion of properties occurred by 38%, associating it with the growth demographic. The above is linked to the increase in urban sprawl, as seen in Fig. 4. This phenomenology was a product of urban expansion in locations south and east of the city. Later, in the 1980s and 1990s, housing growth focused on the southern and southeastern regions. Changes in land use are directly associated with increases in soil temperature.

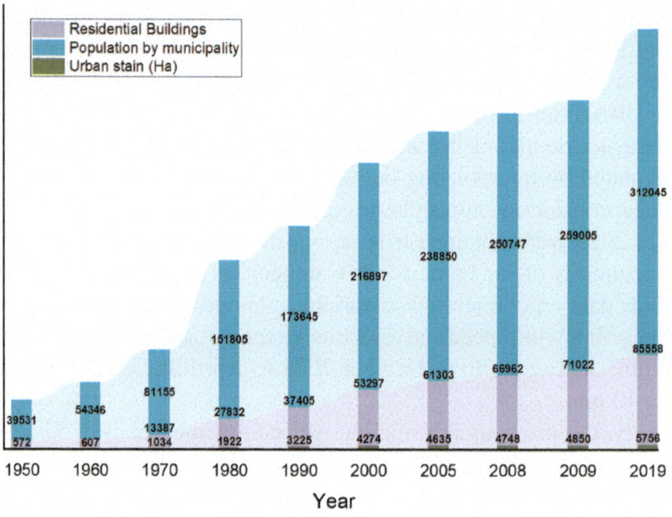

Fig. 4 Urban growth (housing, population, and built area) of the City of San Francisco de Campeche in the last 70 years

4.2 Land Surface Temperature Analysis

The surface temperature was determined by analysing satellite images from thermal bands, which were then processed in geographic information systems (GIS) to visualize the thermal gradients on maps. Figures 5 and 6 show the compilation of LST maps from 1990 to 2020, cataloguing the temperature on the city surface in five colour ranges: blue for temperatures below 20 °C, light blue for values between 20 and 25 °C, green between 20 and 30 °C, yellow between 30 and 35 °C, orange from 35 to 40 °C, and red for temperatures above 40 °C. According to Fig. 5, during the 1990s, the city did not exceed an average of 25 °C at ground level during the hottest time of the year, with the city's oldest and most central neighbourhoods having the highest temperatures. These average temperatures result from a reduced urban area and few buildings, resulting in a high concentration of green areas. However, in subsequent decades until 2020, a significant increase in LST was observed, exceeding 30 °C. It is essential to highlight the expansion of LST in regions surrounding the city due to the urban area's growth. The above agrees with the emergence of housing colonies in the eastern and western regions, causing an expansion of 37% on forest land.

On the other hand, in the regions of the historic centre and the east of the city, highlighted in the figures, the increase in temperature caused by deforestation is observed even exceeding 35 °C (Fig. 6). Furthermore, it is interesting that throughout the graphic record, the temperature of the city is standardized, reducing areas with marked temperature trends below the average. There is even a tendency to generate areas already reaching 40 °C or higher. The above indicates that in 30 years, a coastal city with a small population, such as San Francisco de Campeche, has increased ground-level temperature by approximately ten degrees centigrade.

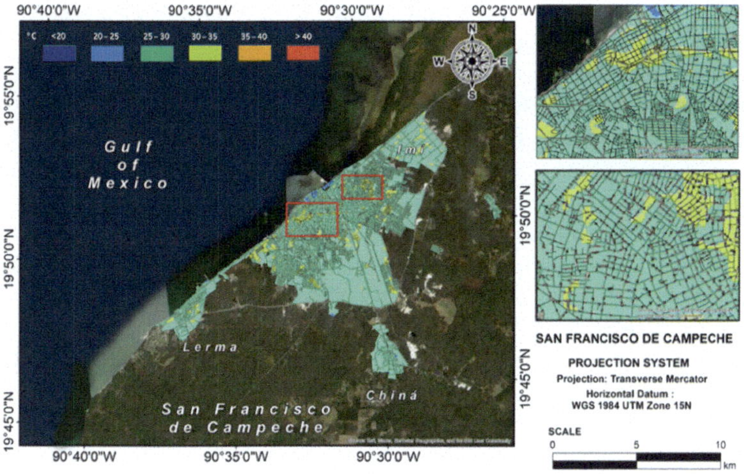

Fig. 5 Surface temperature map for San Francisco de Campeche in 1990

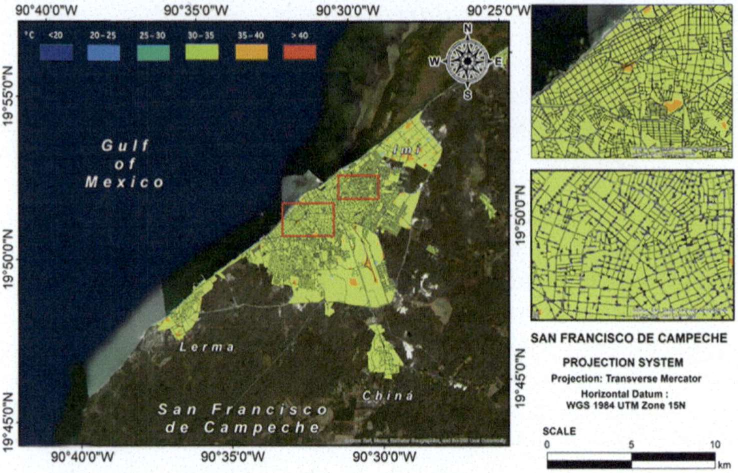

Fig. 6 Surface temperature map for San Francisco de Campeche in 2020

4.3 *Analysis of the Normalized Difference Vegetation Index (NDVI)*

For the analysis of the evolution of vegetation in the city, the NDVI values were grouped into five classes: very sparse vegetation (< 0), sparse vegetation (0–0.25), reduced vegetation (0.25–0.50), acceptable vegetation (0.50–0.75), and abundant vegetation (> 0.75), as seen in Figs. 7 and 8. By 1990, the urban area was small and primarily concentrated in the city centre, characterized by limited vegetation. However, the periphery of the city at that time (now the southern and eastern areas) maintained relatively high levels of vegetation (Fig. 7).

Figure 8 shows that by 2020, the reduction in vegetation had spread to the south and southeast, resulting in more than 90% of the city falling into the sparse vegetation category, contributing to a rise in local temperatures. These findings highlight the crucial role of urban planners, environmental policymakers, and researchers in addressing this issue.

Table 1 lists the evolution of the vegetation index for the five classifications in the last 20 years. According to this, the least frequent category in the city and the one that has been reduced the most has been 'very scarce vegetation (< 0)', which has gone from 16.38 ha to just over 0.38 ha. The same phenomenology occurs in the categories of 'acceptable vegetation' and 'scant vegetation', which have gone from 371.60 to 200.94 ha and 65.25 to 0.37 ha, respectively. The 'scarce vegetation' category has been the most representative and the only one that has grown, going from 181.75 to 433.59 ha, while the 'abundant vegetation range' extensions are almost imperceptible. These results show that over the past two decades, the city has tended to have low levels of vegetation, making it vulnerable to the strong waves of the area's tropical climate due to the absence of plants or trees to mitigate the impact.

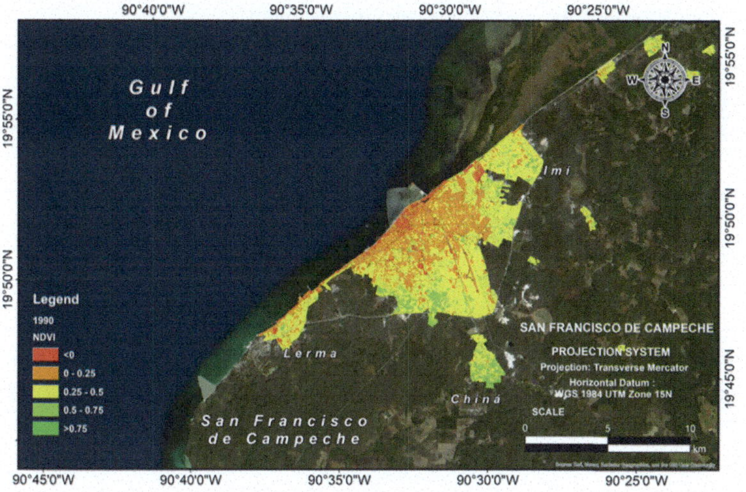

Fig. 7 NDVI map for San Francisco de Campeche in 1990

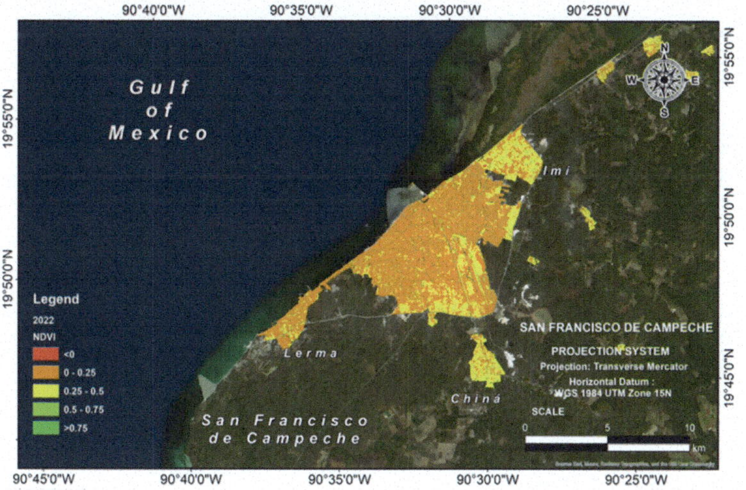

Fig. 8 NDVI map for San Francisco de Campeche in 2022

Regarding Fig. 9, it is established to identify the relationship between the NDVI and LST values for the years under analysis. According to this, between 1990 and 1995, temperatures at ground level did not exceed 35 °C, while the NDVI index was distributed on average between 0.8 and – 0.1, obtaining negative and very steep regression slopes, a sign that the vegetation dampened the thermal effect. Since 2000, there has been a noticeable trend towards more horizontal slopes in the correlation, attributable to rising temperature values (*x*-axis), which rise to 40 °C, and a decrease in

Table 1 Normalized difference vegetation index details from 1990 to 2022

Classes	1990	1995	2000	2015	2020	2022
	Area (ha)	Area (ha)	Area (ha)	Area (ha)	Area (ha)	Area (ha)
< 0.00	116.45	16.52	579.16	0.32	0.14	0.38
0.00–0.25	181.75	386.60	54.27	410.3	515.75	433.59
0.25–0.50	371.60	230.85	1.76	224.4	119.29	200.94
0.5–0.75	65.36	1.22	0.00	0.17	0.01	0.27
> 0.75	0.00	0.00	0.00	0.00	0.00	0.00

vegetation noticeable with average maximum NDVI values around 0.4 and minimum values up to – 0.4. This indicates the presence of UHI in the city is linked to a certain extent to deforestation due to urban sprawl.

4.4 Surface Temperature Time Series

The temperature data were acquired from a location 15 km from the city. This was due to the mesh resolution of the ERAS model, which corresponds to 0.25 °C and provides a data point every 28 km for these operational parameters. The point under analysis is in the coastal area, meaning the data do not ideally characterize the temperature in the city. However, the resolution enables the identification of temperature patterns in the area. Figure 10 exhibits information on temperature behaviour from an annual perspective. This figure demonstrates the impact of thermal warming on temperature behaviour. It shows a steady increase in temperature from 1970 to the present, with maximum temperatures rising in recent decades and minimum temperatures falling as a direct result of thermal warming in the region. Figure 11 presents the temperature series identifying an average temperature of 26.21 °C, with a minimum of 14.01 °C and a maximum value of 36.95 °C, reached in 2020.

5 Discussion of Results

In the preceding sections, the chapter discusses the changes that have taken place in the coastal city of San Francisco de Campeche, which have contributed to the development of the UHI effect. It is evident that the lack of urban design and planning in the city is closely linked to UHIs. Urban areas absorb more solar radiation and have a high thermal capacity due to the presence of buildings, pavements, and various materials and surfaces. This means that urbanization and the expansion of cities have a significant impact on climate and changes in land use. The findings align with those reported by Tian et al. [25], Sultana and Satyanarayan [26], and Han et al. [27], who observed that coastal cities undergo the most significant changes. This is attributed to

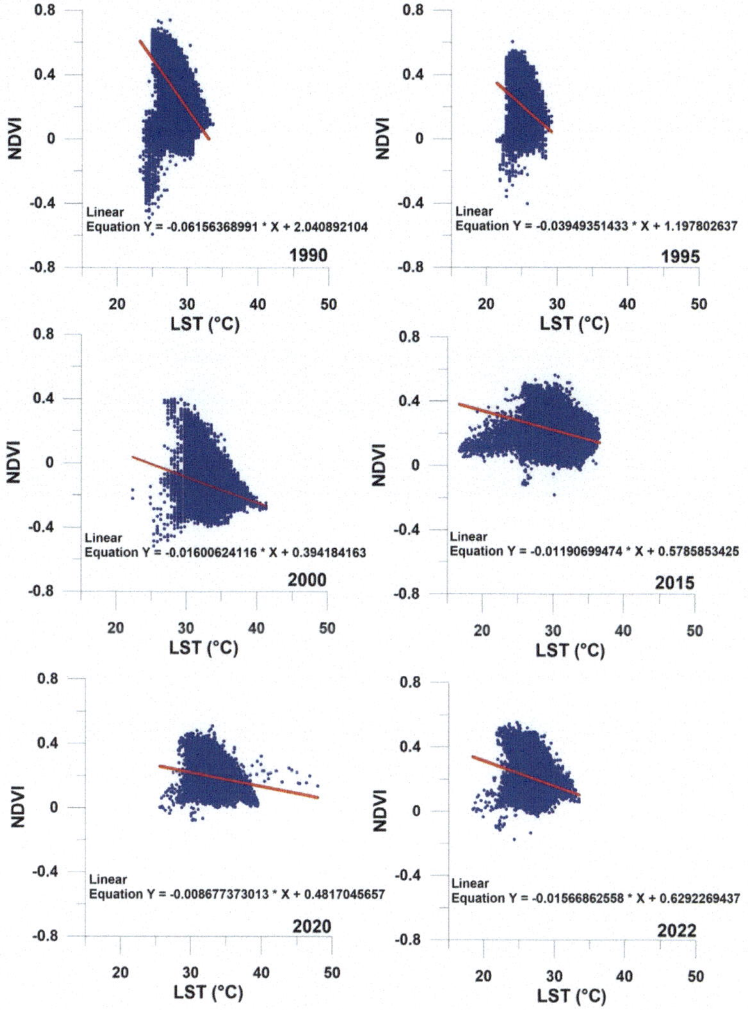

Fig. 9 Relationship between NDVI and LST for San Francisco de Campeche in the years: **a** 1990; **b** 1995, **c** 2000; **d** 2015; **e** 2020; **f** 2022

the rapid expansion of built-up areas in these cities, resulting in a higher proportion, density, and regularity of urban development. Unlike cities with planned expansion that prioritize natural areas and green spaces, coastal cities experience unplanned expansion, contributing to environmental challenges and elevated temperatures.

The UHI phenomenon has adverse effects on the urban social ecosystem, including increased electricity consumption in buildings, reduced thermal comfort, decreased air quality [22], health effects on urban residents [22, 25], and increased mortality. A recent study found that the average annual temperature of a city with more than one million inhabitants is between 1 and 3 °C higher than that of surrounding

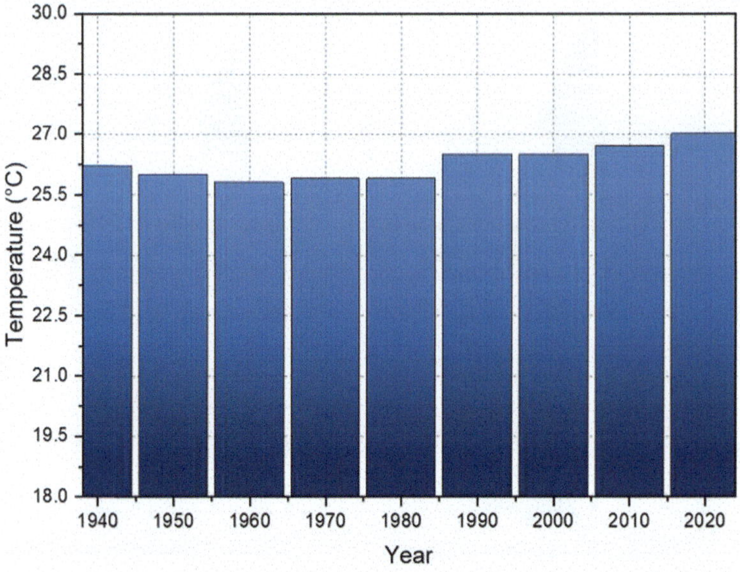

Fig. 10 Annual temperature behaviour from 1940 to 2022 in San Francisco de Campeche

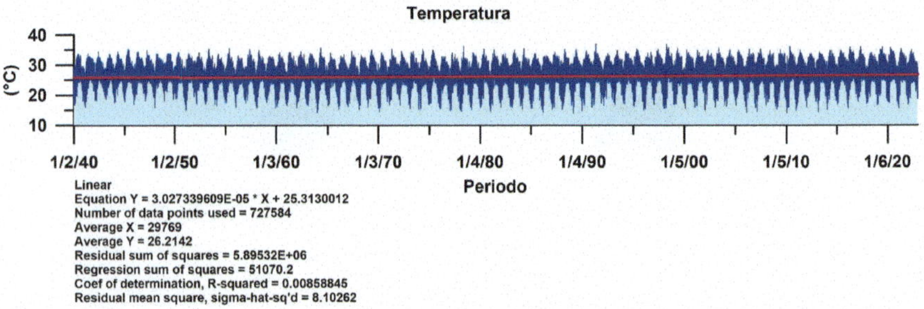

Fig. 11 Hourly temperature behaviour from 1940 to 2022 in San Francisco de Campeche

rural areas [27–29]. When the UHI is intensified, it leads to increased mortality, respiratory diseases, cardiovascular diseases, and even cancer. Changes in land use, coupled with population growth and resulting anthropogenic heat, significantly alter the climate, as shown by [29–31]. Planning and design that modify the features of the surrounding environment could reduce UHIs. Since replacing trees and vegetation with less permeable material surfaces minimizes the natural effects of shade cooling, evaporation of water from the soil, and evapotranspiration of leaves, the reverse process would maximize them [31–33].

Most studies that have examined the UHI effect have found that urbanization leads to changes in the natural landscape and land use, resulting in the loss of vegetation and altering the local environment and climate. NDVI is commonly used to identify these changes. It has proven to be effective in visualizing the areas undergoing the most significant changes in the City of San Francisco de Campeche. Our findings show a relationship between LST and vegetation cover, echoing other research that has observed a positive correlation between LST, pollution, and changes in land cover in urban areas, accompanied by a negative correlation between LST and green areas [32, 33].

Due to changes driven largely by the construction sector, which accounts for 27% of global greenhouse gas emissions, various strategies have been proposed and implemented worldwide to mitigate the UHI effect. These strategies include creating urban green spaces, green roofs, vertical greening, bodies of water, using cold materials, and changes in urban design. Studies have focused on the impact of these strategies on reducing UHI effects, energy consumption in buildings, and outdoor thermal comfort. Sustainable urban infrastructure, effective rainfall management, and reduced human-induced heat have also been suggested. Mitigation measures in construction involve protection from solar radiation, minimizing heat infiltration, reducing human-induced heat, maintaining thermal comfort, and urban area planning. Urban development measures include reforestation, green infrastructure, and reducing human-induced heat. In addition, public policies, certifications, and regulations should be developed, such as the European Optimal Profitability Methodology [22, 27, 33, 34], proposed as a guiding method to outline the most cost-effective measures to reconstruct existing buildings, focusing on economic aspects. As a result, retrofit measures that integrate green strategies at the building level (i.e. green roofs or walls) or interventions to achieve a nearly zero-energy buildings (NZEB) standard are penalized by their high upfront investment costs, as the environmental benefits are not adequately assessed. As for energy and environmental certifications, they assign points, mainly referring to the minimization of heat island effects, both in the urban area and at the construction level, or as Italian standard which regulates the development of the urban environment to comply with the Kyoto Protocols, highlighting the role of trees [33].

In San Francisco de Campeche, there has been an observed increase in temperature over the last 30 years. Although not addressed in this study, it is assumed that electrical energy consumption has also increased. Tian et al. [25] found that in countries with hot climates, a 1 °C increase in temperature leads to a 1.66% increase in electricity consumption. Additionally, there is a link between increased electricity consumption, UHI-related diseases, and urban thermal environment. Implementing the abovementioned strategies could significantly impact the urban thermal environment, mainly if used during the project's design stage. For example, urban greening can purify the air, regulate local temperature or climate, and improve the urban ecosystem; urban green space allows to reduce air temperature and mitigates its pollution, as well as reduces energy consumption for cooling; on the other hand, green roofs can influence the urban environment, since they represent between 20 and 25% of the surface of a city, which is how Besir and Cuce [36] found that they can reduce indoor temperatures on the top floor by up to 3.4 °C. In Hong Kong, a study demonstrated that achieving 60% green cover within communities could lower air temperatures by 0.65–1.45 °C [37]. This initiative also estimated annual energy

Fig. 12 Intervention proposal for the urban sector of the city of San Francisco de Campeche: current status of the urban sector (top images); proposal for green intervention in the urban sector (bottom images)

savings ranging from 3.4×10^7 to 7.6×10^7 kWh. Green spaces not only improve aesthetics but also effectively mitigate the urban heat island effect by lowering wall temperatures, offering thermal insulation, cooling through evapotranspiration, and serving as windbreaks. Research has shown that installing green walls can reduce a building's energy consumption by up to 16%. Figure 12 illustrates the before-and-after results of a reforestation proposal in an urban sector and the installation of a green wall on housing [38, 39].

6 Conclusions

It is essential to highlight the approach to studying heat islands in San Francisco de Campeche, as this research complements traditional climate and weather studies. These studies should be integrated with urban planning and city design observations, notably since the city has expanded westwards in recent years, as shown in satellite images. This integration is crucial for comprehensively understanding how urban growth impacts local climate conditions and helps devise effective mitigation strategies.

The heat islands identified and analysed in the study correspond to the areas with the greatest urban infrastructure in the city, taking as a reference area where the largest number of buildings are concentrated, a significant number of vehicles, and the smallest area of vegetation or the scarcity of it. By focusing on these factors, locations where heat islands are most prominent can be pinpointed, thereby providing critical data for urban planning and environmental mitigation strategies.

The points analysed demonstrate the spatial relationship of temperature increases from 1940 to 2020, with an approximate rise of 6 °C between 1990 and 2022. This temperature increase corresponds to the decrease in vegetation due to urban sprawl and the rise in housing developments, as evidenced by the quadrupling of the NDVI values in the 0–0.25 range during the same period. The annual maximum temperature values series study indicates a trend of increasing temperatures.

Finally, to mitigate heat islands and improve the quality of the city's habitat, it is essential to evaluate strategies and public policies that could be implemented in the city. These could range from trees with local vegetation to vertical greening or green walls to regulate air temperature or local climate, reduce energy consumption, and improve the urban ecosystem.

Acknowledgements This work is a product of the Climatological Observatory of San Francisco de Campeche KEY: 036/UAC/2023 corresponding to the institutional research group UNACAM-CA-56. The authors also acknowledge the support provided by the Thematic Network 722RT0135 '*Red Iberoamericana de Pobreza Energética y Bienestar Ambiental*' (RIPEBA) financed by the call for Thematic Networks of the CYTED Program for 2021.

References

1. Jiménez Torres M, Bienvenido-Huertas D, May Tzuc O et al (2023) Assessment of climate change's impact on energy demand in Mexican buildings: projection in single-family houses based on representative concentration pathways. Energy Sustain Dev 72:185–201. https://doi.org/10.1016/j.esd.2022.12.012
2. Kasniza Jumari NAS, Ahmed AN, Huang YF et al (2023) Analysis of urban heat islands with landsat satellite images and GIS in Kuala Lumpur Metropolitan City. Heliyon 9:e18424. https://doi.org/10.1016/j.heliyon.2023.e18424
3. Khare VR, Vajpai A, Gupta D (2021) A big picture of urban heat island mitigation strategies and recommendation for India. Urban Clim 37:100845. https://doi.org/10.1016/j.uclim.2021.100845
4. Ortiz Porangaba GF, Teixeira DCF, Amorim MCCT et al (2021) Modeling the urban heat island at a winter event in Três Lagoas, Brazil. Urban Clim 37:100853. https://doi.org/10.1016/j.uclim.2021.100853
5. Kim SW, Brown RD (2021) Urban heat island (UHI) intensity and magnitude estimations: a systematic literature review. Sci Total Environ 779:146389. https://doi.org/10.1016/j.scitotenv.2021.146389
6. Singh N, Singh S, Mall RK (2020) Urban ecology and human health: implications of urban heat island, air pollution and climate change nexus. In: Urban ecology. Elsevier, pp 317–334

7. OPS, OMS (2016) Calidad del Aire Ambiente. In: Organ. Panam. la Salud. https://www.who.int/es/news-room/fact-sheets/detail/ambient-(outdoor)-air-quality-and-health. Accessed 8 Aug 2023
8. Vasenev V, Varentsov M, Konstantinov P et al (2021) Projecting urban heat island effect on the spatial-temporal variation of microbial respiration in urban soils of Moscow megalopolis. Sci Total Environ 786:147457. https://doi.org/10.1016/j.scitotenv.2021.147457
9. Xu L, Wang J, Xiao F et al (2021) Potential strategies to mitigate the heat island impacts of highway pavement on megacities with considerations of energy uses. Appl Energy 281:116077. https://doi.org/10.1016/j.apenergy.2020.116077
10. Ma Y, Lauwaet D, Kouti A, Verbeke S (2023) A toolchain to evaluate the impact of urban heat island and climate change on summer overheating at district level. Urban Clim 51:101602. https://doi.org/10.1016/j.uclim.2023.101602
11. Liang Z, Huang J, Wang Y et al (2021) The mediating effect of air pollution in the impacts of urban form on night time urban heat island intensity. Sustain Cities Soc 74:102985. https://doi.org/10.1016/j.scs.2021.102985
12. Luo Y, He J (2021) Evaluating the heat island effect in a planned residential area using planning indicators. J Build Eng 43:102473. https://doi.org/10.1016/j.jobe.2021.102473
13. Qiu J, Li X, Qian W (2023) Optimizing the spatial pattern of the cold island to mitigate the urban heat island effect. Ecol Indic 154:110550. https://doi.org/10.1016/j.ecolind.2023.110550
14. Freitas ED, Rozoff CM, Cotton WR, Dias PLS (2007) Interactions of an urban heat island and sea-breeze circulations during winter over the metropolitan area of São Paulo, Brazil. Boundary-Layer Meteorol 122:43–65. https://doi.org/10.1007/s10546-006-9091-3
15. Yun GY, Ngarambe J, Duhirwe PN et al (2020) Predicting the magnitude and the characteristics of the urban heat island in coastal cities in the proximity of desert landforms. The case of Sydney. Sci Total Environ 709:136068. https://doi.org/10.1016/j.scitotenv.2019.136068
16. Degirmenci K, Desouza KC, Fieuw W et al (2021) Understanding policy and technology responses in mitigating urban heat islands: a literature review and directions for future research. Sustain Cities Soc 70:102873. https://doi.org/10.1016/j.scs.2021.102873
17. INEGI IN de E y G (2022) Extensión Territorial. Cuéntame de México. https://cuentame.inegi.org.mx/territorio/extension/default.aspx?tema=T. Accessed 12 June 2022
18. May Tzuc O, Jiménez Torres M, Rodriguez CM et al (2023) Cluster analysis as a tool for the territorial categorization of energy consumption in buildings based on weather patterns, pp 73–91
19. Nava Fuentes JC, Arenas Granados P, Cardoso Martins F (2018) Integrated coastal management in Campeche, Mexico; a review after the Mexican marine and coastal national policy. Ocean Coast Manag 154:34–45. https://doi.org/10.1016/j.ocecoaman.2017.12.029
20. INEGI; Instituto Nacional de Estadística y Geografía Clima. Campeche. https://cuentame.inegi.org.mx/monografias/informacion/camp/territorio/clima.aspx?tema=me&e=04. Accessed 25 June 2024
21. INEGI; Instituto Nacional de Estadística y Geografía Número de habitantes. Campeche. https://cuentame.inegi.org.mx/monografias/informacion/camp/poblacion/. Accessed 25 June 2024
22. Li X, Li W, Middel A et al (2016) Remote sensing of the surface urban heat island and land architecture in Phoenix, Arizona: combined effects of land composition and configuration and cadastral–demographic–economic factors. Remote Sens Environ 174:233–243. https://doi.org/10.1016/j.rse.2015.12.022
23. Wang R, Hou H, Murayama Y, Derdouri A (2020) Spatiotemporal analysis of land use/cover patterns and their relationship with land surface temperature in Nanjing, China. Remote Sens 12:440. https://doi.org/10.3390/rs12030440
24. ECMWF (2021) Copernicus climate change service. https://cds.climate.copernicus.eu/api-how-to. Accessed 4 July 2024
25. Tian L, Li Y, Lu J, Wang J (2021) Review on urban heat island in China: methods, its impact on buildings energy demand and mitigation strategies. Sustainability 13:762. https://doi.org/10.3390/su13020762

26. Sultana S, Satyanarayana ANV (2020) Assessment of urbanisation and urban heat island intensities using landsat imageries during 2000–2018 over a sub-tropical Indian City. Sustain Cities Soc 52:101846. https://doi.org/10.1016/j.scs.2019.101846
27. Han W, Tao Z, Li Z et al (2022) Effect of urban built-up area expansion on the urban heat islands in different seasons in 34 metropolitan regions across China. Remote Sens 15:248. https://doi.org/10.3390/rs15010248
28. Leal Filho W, Echevarria Icaza L, Emanche V, Quasem Al-Amin A (2017) An evidence-based review of impacts, strategies and tools to mitigate urban heat islands. Int J Environ Res Public Health 14:1600. https://doi.org/10.3390/ijerph14121600
29. Shu C, Gaur A, Wang L, Lacasse MA (2023) Evolution of the local climate in Montreal and Ottawa before, during and after a heatwave and the effects on urban heat islands. Sci Total Environ 890:164497. https://doi.org/10.1016/j.scitotenv.2023.164497
30. Khamchiangta D, Dhakal S (2019) Physical and non-physical factors driving urban heat island: case of Bangkok metropolitan administration, Thailand. J Environ Manage 248:109285. https://doi.org/10.1016/j.jenvman.2019.109285
31. Hidalgo-García D, Arco-Díaz J (2023) Spatiotemporal analysis of the surface urban heat island (SUHI), air pollution and disease pattern: an applied study on the city of Granada (Spain). Environ Sci Pollut Res 30:57617–57637. https://doi.org/10.1007/s11356-023-26564-7
32. Pandey D, Tiwari KC (2020) Extraction of urban built-up surfaces and its subclasses using existing built-up indices with separability analysis of spectrally mixed classes in AVIRIS-NG imagery. Adv Sp Res 66:1829–1845. https://doi.org/10.1016/j.asr.2020.06.038
33. Valdiviezo-N JC, Téllez-Quiñones A, Salazar-Garibay A, López-Caloca AA (2018) Built-up index methods and their applications for urban extraction from Sentinel 2A satellite data: discussion. J Opt Soc Am A 35:35. https://doi.org/10.1364/JOSAA.35.000035
34. Zhao W, Zhu X, Jiang G, Gao Z (2018) Seasonal variation and land-use/land-cover type impacts on the correlation of urban heat island intensity and difference vegetation index with satellite data in Xi'an, China. Wuhan Univ J Nat Sci 23:387–395. https://doi.org/10.1007/s11859-018-1338-6
35. Hu Y, Dai Z, Guldmann J-M (2020) Modeling the impact of 2D/3D urban indicators on the urban heat island over different seasons: a boosted regression tree approach. J Environ Manage 266:110424. https://doi.org/10.1016/j.jenvman.2020.110424
36. Besir AB, Cuce E (2018) Green roofs and facades: a comprehensive review. Renew Sustain Energy Rev 82:915–939. https://doi.org/10.1016/j.rser.2017.09.106
37. Tam VWY, Wang J, Le KN (2016) Thermal insulation and cost effectiveness of green-roof systems: an empirical study in Hong Kong. Build Environ 110:46–54. https://doi.org/10.1016/j.buildenv.2016.09.032
38. Ciacci C, Banti N, Di Naso V et al (2022) Experimentation of mitigation strategies to contrast the urban heat island effect: a case study of an industrial district in italy to implement environmental codes. Atmosphere 13:1808. https://doi.org/10.3390/atmos13111808
39. Union EP and C of the E Directive—2010/31—EN—EUR-Lex. https://eur-lex.europa.eu/legal-content/EN/ALL/?uri=celex%3A32010L0031. Accessed 26 June 2024

The Influence of Global Warming and Tourism Over the Sleeping Thermal Comfort of Dwellings in Granada

Antonio García Molina, María Luisa de la Hoz Torres, Antonio Jesús Aguilar, Beatriz Nestares Nieto, and David Bienvenido Huertas

Abstract The impact of tourism on the behavioural patterns of residential occupants can be significant. Noise generated by tourist activities may restrict the resident ability of adaptation during hot periods (e.g. limiting opening windows). This issue can be even more severe in future climate change scenarios. In this context, the aim of this study is to analyse the impact of restricted natural ventilation practices on sleeping thermal comfort. The analysis focused on the summer period (e.g. July and August) when natural nocturnal ventilation can effectively cool residential spaces. A case study of a dwelling in the Albaicín neighbourhood of Granada was selected for this purpose. Simulations were conducted for both current and future scenarios based on representative concentration pathways (RCP 2.6, RCP 4.5, and RCP 8.5). The thermal comfort of the dwelling was assessed using the PMV model adapted for the night-time conditions. The findings indicate that limiting natural night ventilation leads to an increase in hours of thermal discomfort. For all future scenarios, the values obtained indicate that the percentage of hours of sleeping thermal discomfort is higher compared to the current scenario.

A. García Molina
University of Granada, Granada, Spain

M. L. de la Hoz Torres
Department of Building Construction II, University of Seville, Seville, Spain

A. J. Aguilar
University of Seville, Seville, Spain

B. Nestares Nieto
Division of Civil Engineering and Built Environment, Department of Civil and Industrial Engineering, Uppsala University, Uppsala, Sweden

D. Bienvenido Huertas
Department of Building Construction, University of Granada, Granada, Spain

M. L. de la Hoz Torres (✉) · A. J. Aguilar
Department of Signal Theory and Communications, Rey Juan Carlos University, Fuenlabrada, Spain
e-mail: maria.delahoz@urjc.es

M. Torres González and C. Rubio Bellido (eds.), *Towards Low and Positive Energy Buildings*, Green Energy and Technology,
https://doi.org/10.1007/978-3-031-70851-0_12

1 Introduction

Tourist activities can have a significant negative impact over the lifestyle of permanent city residents. The transformation of residential neighbourhoods to adjust to visitors needs can lead to various problems, including the gentrification and multiple coexistence conflicts for local residents. One of the problems associated with this high level of tourism is noise pollution, particularly during night-time. The noise generated by tourism activities occurs throughout the year and seriously degrades the habitability and comfort conditions of dwellings. This noise pollution can also lead to health issues, such as stress, disrupted sleep, and fatigue [1, 2]. As a result, residents of neighbourhoods with high tourist occupancy are forced to isolate themselves in the interior of their homes, keeping windows and external openings normally closed. This situation affects the indoor thermal comfort conditions in the buildings, limiting the personal adaptive behaviours of occupants compared to naturally ventilated buildings [3–5] and triggering the increase used of air conditioning systems, which directly impacts the amount of energy resource requirements [6].

These factors converge with the gradual increase of outdoor temperatures due to climate change. Climate change and environmental degradation may soon reach critical levels, especially due to ongoing emissions of greenhouse gases (GHGs). Addressing the root causes of these emissions is a global priority. In this context, the 2015 Paris Agreement committed countries to reduce emissions. For now, however, the goals set out in the deal are far from being achieved [7]. In any case, these targets established a requirement that all human activities would result in zero GHG emissions. One of the major activities generating significant emissions is the use of buildings, which accounts for 36% of GHG emissions. As a result, the European Union has set a goal of reducing greenhouse gas emissions by 90% by 2050 [8].

Most scientific and engineering efforts in the field have focused on improving passive (envelope) and active (system) components of the building [9–12]. However, the role of users in managing GHG emission producing activities such as HVAC systems is often overlooked. Previous studies have emphasized the benefits of using adaptative strategies to reduce energy consumption, such as natural ventilation, extended thermal comfort range in thermostat settings, and employing free-running building regimes [13–17]. These strategies could achieve a significant reduction of the energy demand in buildings.

However, while the incorporation of these strategies has been extensively studied for day-time, few studies have investigated the consequences of the limited applications these studies have for night-time and their implications for thermal comfort. During night-time hours, outdoor temperature is lower than during the day, allowing significant energy savings through natural ventilation [18, 19].

Limiting natural ventilation during the night has a critical impact on residents, as thermal comfort is directly related to sleep quality [20–22]. Wang et al. found that even slight deviations from optimal thermal conditions can lead to poorer sleep quality among surveyed users [20]. Moreover, heat exposure during night hours can disrupt rest more significantly than cold exposure [21]. Deviations from sleeping thermal comfort conditions can result in reduced sleep quality, leading to effects such as decreased daytime performance [22] and overall health deterioration [23]. Therefore, it is crucial to analyse sleeping thermal comfort ventilation, as users' adaptability ranges differ from those during the day [24].

Despite these considerations, most studies evaluating the effect of thermal comfort over the sleep quality have been focused primarily on cold seasons [25, 26]. In general, there are limited studies that research sleeping thermal comfort during warm seasons, particularly in the context of climate change. Therefore, further research is essential to evaluate the impact of nocturnal indoor thermal conditions on occupant comfort. This study aims to address this gap.

In this context, the pressure of tourism on residential historic districts can limit the possibility of natural ventilation of buildings during summer nights, affecting negatively to the residents' quality of life. The main goal of this book chapter is to analyse sleeping thermal comfort in a dwelling located in a historic district of Granada, considering both current and future RCP scenarios according to the IPCC panel and determinate the influence and effects that global warming has for these conditions.

2 Methodology

2.1 Case Study

For this study, a characteristic residential building in the historic district of Granada, Spain, was selected. Granada is one of the most touristic cities in Spain, and its residential neighbourhoods are under significant pressure as a result from this activity [27]. Granada is characterized by a hot-summer Mediterranean climate (Csa) according to the Koppen Gueiguer classification, with hot and dry summers (temperatures averaging 34 ºC during the hottest month). Nevertheless, high outdoor temperatures are commom during warm season, reaching values over 40 ºC. During night, the mean minimum outdoor temperatures range between 18 and 25 ºC [28].

The study focused on a first-floor apartment featuring a total area of 72 m^2 and comprising two east facing bedrooms (Fig. 1). The thermal transmittance (U-factor) of building envelope systems is 0.566 W/m^2K. There is no air conditioning system in any of the rooms. Prior interviews with the residents of the selected dwelling indicated that they experience significant thermal discomfort during summer nights. This discomfort is attributed to the necessity of keeping windows closed due to external noise, coupled with the avoidance of air conditioning because of high energy costs.

After selecting the building for the case study, a model was develop using Design-Builder software (version 7) to perform time series simulations of indoor environmental variables under both current and future scenarios (RCP 2.6, RCP 4.5, and

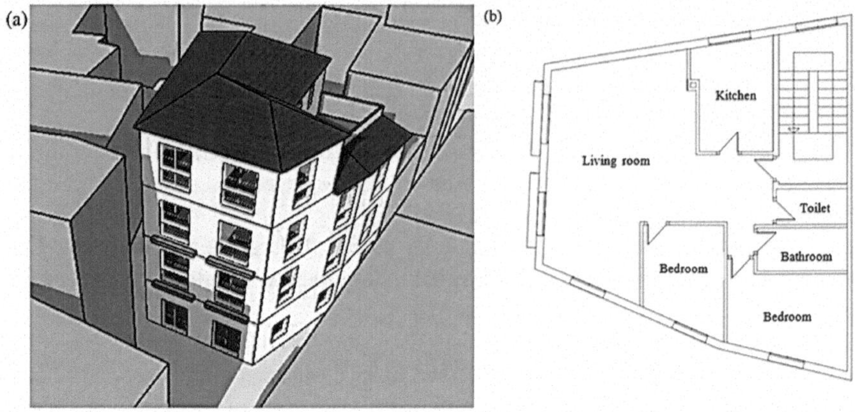

Fig. 1 **a** Geometry of the building used in the case study and **b** planimetry of the first floor

RCP 8.5). The building model was validated according to the criteria defined in ASHRAE Guideline 14-2014 [29].

2.2 *Sleeping Thermal Comfort Model*

The thermal comfort procedure used in the study is based in the adaptation of the method suggested by Fanger. This method relies on the predicted mean vote (PMV), which stablished the relationship between the individuals' thermal sensation and the thermal environment based on experimental research conducted in controlled climatic settings [30]. However, this method has limitations as it is designed for activities involving some level of physical activity. Consequently, applying the model does not yield accurate results for sleeping individuals. Fanger's PMV model was defined for a metabolic rate range between 0.8 and 4 MET, whereas the metabolic rate during sleep is much lower (approximately 0.7 MET) [31].

To address these issues, researchers including Lan et al. have created an adjusted version of Fanger's model to better assess sleeping thermal comfort [32]. This model considers elements such as skin temperature fluctuations, decreased heat loss from respiration during sleep, and the insulation properties of the bedding materials (mattress, sheets, etc.). Equation 1 illustrates the calculation method used to estimate the thermal load of one person sleeping, which is the difference between the internal heat production and the heat loss to the indoor environment.

$$L = M - Q_{\text{res}} - Q_{\text{skin}} \tag{1}$$

where M is the heat generated by the user while sleeping, Q_{res} represents the heat loss through breathing, and Q_{skin} represents the heat loss through the skin. The Q_{res} can be calculated as the sum of latent and sensible heat losses, which can be estimated using Eq. 2.

$$Q_{\text{res}} = \frac{13.41 - 1.519 \times 10^{-3} p_a - 0.13 t_a}{A_D} \tag{2}$$

where p_a is the partial pressure of water vapour present in the inspired air (Pa), t_a is the temperature of the inspired air (ºC), and A_D is the uncovered surface area of the body (m^2). A_D can be obtained using Eq. 3 for men and Eq. 4 for women, based on the weight (w) and height (l) of the individual being analysed.

$$\text{Men} \rightarrow A_D = 0.607 \cdot l + 0.0127 \cdot w - 0.0698 \tag{3}$$

$$\text{Women} \rightarrow A_D = 0.589 \cdot l + 0.0126 \cdot w - 0.0461 \tag{4}$$

According to this model, the heat flow between the body and the mattress occurs through conduction and can be estimated using Eq. 5.

$$E_{\text{cond}} = \alpha k \frac{t_{\text{sk2}} - t_b}{d} \tag{5}$$

where E_{cond} is the heat loss by conduction between the body and the mattress (W/m^2), α is the percentage of the body part in contact with the mattress, k is the thermal conductivity coefficient of the mattress (W/m·K), t_{sk2} is the average skin temperature of the body part in contact with the mattress (°C), t_b is the surface temperature of the mattress (°C), and d is the thickness of the mattress (m).

The heat loss by skin diffusion (E_d) is considered to decrease by 20% if the body is not in contact with the bed and is calculated using Eq. 6.

$$E_d = 3074 \times 10^{-3} (1 - 0.8\alpha)(p_{\text{sk1}} - p_a) \tag{6}$$

where p_{sk1} is the saturated vapour pressure at the average skin temperature of the body parts that are not in contact with the mattress (p_a). In addition to this loss, there is also heat loss by convection and radiation, which is determined using Eq. 7.

$$E_{C+R} = (1 - \alpha) \frac{t_{\text{piel1}} - t_o}{R_{cl} + \frac{1}{f_{cl}(h_r + h_c)}} \tag{7}$$

where t_{sk1} is the average skin temperature of the body parts that are not in contact with the mattress (°C), t_o is the operative temperature (°C), h_c is the convective heat transfer coefficient ($W/m^2 \cdot K$), h_r is the radiative heat transfer coefficient ($W/m^2 \cdot K$), f_{cl} is the covering area factor, and R_{cl} is the thermal resistance of the sleep covering (m^2 K/W). For a person in the supine position, a value of h_r of 3.235 $W/m^2 \cdot K$ can be assumed, while h_c can be calculated using Eq. 8 [33].

$$h_c = 0.881|t_{\mathrm{sk1}} - t_a|^{0.368} \tag{8}$$

Therefore, based on the aforementioned considerations, the thermal load of a sleeping person can be estimated using Eq. 9.

$$L = 40 - \frac{13.41 - 1.519 \times 10^{-3} p_a - 0.13 t_a}{A_D} - X \tag{9}$$

where X can be calculated thanks to Eq. 10:

$$\begin{aligned} X &= 3.074 \times 10^{-3}(1-\alpha)(p_{\mathrm{sk1}} - p_a) \\ &+ (1-\alpha)\frac{t_{\mathrm{sk1}} - t_o}{R_{cl} + \frac{1}{f_{cl}(h_r + h_c)}} + \alpha k \frac{t_{\mathrm{sk2}} - t_a}{d} \end{aligned} \tag{10}$$

For the application of this model, the following considerations were assumed: (i) the user of the dwelling is 1.82 m tall and weighs 99 kg; (ii) the mattress in the master bedroom is made of latex with an assumed thermal conductivity of 0.048 W/m^2 K and a thickness of 0.20 m; (iii) the metabolic rate was assumed to be 0.7 Met, characteristic of sleeping individuals [34]; and (iv) p_{sk1} was considered to be 5520 Pa, t_{sk1} to be 34.6 °C, and t_{sk2} to be 35.4 °C, according to Lan et al. [32]. Once L is determined, PMV can be calculated using Eq. 11:

$$\mathrm{PMV} = \left(0.303 e^{-0.036M} + 0.028\right) L \tag{11}$$

The thermal comfort ranges considered in this study are those established for existing buildings by the EN 16798-1 standard [35]. The thermal comfort range was defined as PMV values between – 0.70 and 0.70. The analysis was conducted during the night-time hours of July and August, specifically between 11:00 pm and 8:59 am.

2.3 Considered Climate Scenarios

For this purpose, 3 RCP scenarios have been considered with different levels of climate worsening: RCP 2.6 (lowest level of worsening with an increase of between 0.3 and 1.7 °C in the global average temperature at the end of the twenty-first century), RCP 4.5 and RCP 8.5 (highest level of worsening, with an increase of between 1.1 and 2.6 °C for RCP 4.5 and between 2.6 and 4.8 °C for RCP 8.5).

Given that the dwelling is located in a high-tourist-pressure area of Granada, it was assumed that the occupants could not open windows for natural ventilation during the night. This limitation was taken into account during the development of simulations for each scenario. Additionally, the future scenarios were evaluated for both 2050 and 2100 to understand the expected thermal behaviour inside the dwelling throughout the twenty-first century.

3 Results and Discussion

As indicated in the methodology section, all thermal comfort results were obtained according to the situation and operational profile of the dwelling's occupants (i.e. sleeping with the windows closed to avoid noise issues). This results in higher indoor temperatures during summer nights compared to the outdoor temperatures, as natural ventilation is not possible. Table 1 provides a summary of the variations in indoor temperature for the months of July and August in each of the scenarios.

$\overline{T_{air-in}}$ mean indoor air temperature; $\overline{\Delta T}$ mean indoor temperature increase compared to outdoor temperature; $T_{air,in}$ indoor air temperature obtained from simulations.

The analysis reveals an increasing trend in the indoor temperature of dwellings in the different scenarios. Noise pollution and the limited implementation of natural ventilation strategies impact the thermal comfort of the occupants. The results imply that if natural ventilation through windows was possible, indoor temperatures would be 4.66, 3.31, and 1.17 °C lower in the current situation (2024), in the RCP 8.5 scenario of 2050, and in the RCP 8.5 scenario of 2100, respectively. Consequently, the percentage increase in indoor temperature due to touristification is 19.36% in the current scenario, decreasing to 12.97% in the RCP 8.5 scenario of 2050 and to 4.23% in the RCP 8.5 scenario of 2100.

These variations in indoor temperature result in fluctuations in the percentage of sleeping thermal comfort hours (Table 2).

Table 1 Table of night-time temperature increases due to lack of ventilation caused by noise pollution

Scenario		$\overline{T_{air-in}}$ July/August (°C)	$\overline{T_{air,\,in}}$ summer (°C)	$\overline{\Delta T}$ July/August (°C)	$\overline{\Delta T}$ summer (°C)	Percentage of $T_{air,in}$ due to Touristification (%)
2024	Current situation	24.14/24.01	24.07	4.60/4.72	4.66	19.36
2050	RCP 2.6	24.97/24.83	24.90	3.80/3.90	3.85	15.46
	RCP 4.5	25.28/25.14	25.21	3.54/3.63	3.59	14.24
	RCP 8.5	25.66/25.40	25.53	3.20/3.41	3.31	12.97
2100	RCP 2.6	25.02/24.80	24.91	3.78/3.92	3.85	15.46
	RCP 4.5	26.13/25.90	26.02	2.82/3.03	2.93	11.26
	RCP 8.5	27.77/27.52	27.65	1.08/1.26	1.17	4.23

Table 2 Summary of sleeping comfort hours (NCH) and sleeping discomfort hours (NDH) in the various scenarios studied

Year	Scenario	NCH July (h)	NCH August (h)	Total NCH (h)	Total NCH (%)	Total NDH (h)	Total NDH (%)
2024	Current situation	214	200	414	66.77	206	33.23
2050	RCP 2.6	142	110	252	40.65	368	59.35
	RCP 4.5	117	83	200	32.26	420	67.74
	RCP 8.5	69	53	122	19.68	498	80.32
2100	RCP 2.6	122	116	238	38.39	382	61.61
	RCP 4.5	37	53	90	14.52	530	85.48
	RCP 8.5	0	0	0	0.00	620	100.00

The analysis of the number of sleeping thermal comfort hours in the dwelling shows that 66.77% of the hours are comfortable in the current scenario. However, this value decreases due to the impact of climate change in the different scenarios analysed, reaching 19.68% of comfortable hours in 2050 (RCP 8.5) and 0% in 2100 (RCP 8.5).

In the RCP 2.6 scenario, it is observed that the 66.77% of thermally comfortable nocturnal hours in the current scenario drops to 40.65% in 2050 and to 38.39% in 2100. This means that by the end of the century, there is expected to be 1.85 times more discomfort hours compared to the current scenario. In the RCP 4.5 scenario, the percentage of thermal comfort hours is 32.26% in 2050 and 14.52% in 2100. This results in an increase in the percentage of discomfort hours by 2.57 times. In the RCP 8.5 scenario, 19.68% of thermal comfort hours are obtained in 2050 and 0.00% in 2100.

Therefore, the results show that in future scenarios, the percentage of sleeping thermal comfort hours will decrease. This percentage is affected by the limitation in adopting nocturnal natural ventilation strategies. Consequently, the high tourist pressure and the resulting noise pollution worsen the thermal comfort of the occupants of these dwellings in the current scenario and even more severely in future scenarios.

Tables 3 and 4 show the results obtained after applying the PMV model adapted to the nocturnal period in July and August. For July, the results indicate that the mean PMV value ($\overline{PMV}$) is 0.51 in the current scenario. However, it is important to note that a maximum PMV value (PMV_{max}) of 1.32 is reached, which exceeds the upper limit of thermal comfort ($PMV = 0.7$).

For the year 2050, the different scenarios show an increase in the $\overline{PMV}$, reaching values of 0.76 (RCP 2.6), 0.89 (RCP 4.5) and 0.99 (RCP 8.5). These values indicate that in all 2050 scenarios, the $\overline{PMV}$ is above the comfort limit ($PMV = 0.7$). This means that the mean $\overline{PMV}$ is 8.57%, 27.14%, and 41.43% above the comfort limit in the RCP 2.6, RCP 4.5, and RCP 8.5 scenarios, respectively.

A similar trend is observed in the results for July of the year 2100, where the $\overline{PMV}$ reaches values of 0.80, 1.17, and 1.73 in the RCP 2.6, RCP 4.5, and RCP

Table 3 Summary of sleeping thermal comfort PMV values for the case study in July

Scenario		Q	PMV ranges	$\overline{PMV}$	PMV_{max}	Dist. PMV to CV_{max}	DI (%)	Dist. $\overline{PMV}$ to CV_{max}	DIMV
2024		I	– 0.35–0.23	0.51	1.32	–	–	–	–
		II	0.23–0.50			–	–		
		III	0.50–0.80			0.10	14.29		
		IV	0.80–1.32			0.62	88.57		
2050	RCP 2.6	I	-0.07–0.52	0.76	1.75	–	–	0.06	8.57
		II	0.52–0.77			0.07	10.00		
		III	0.77–0.99			0.29	41.43		
		IV	0.99–1.75			1.05	150.00		
	RCP 4.5	I	– 0.05–0.57	0.89	1.96	–	–	0.19	27.14
		II	0.57–0.85			0.15	21.43		
		III	0.85–1.18			0.48	68.57		
		IV	1.18–1.96			1.26	180.00		
	RCP 8.5	I	0.14–0.74	0.99	1.94	0.04	5.71	0.29	41.43
		II	0.74–1.03			0.33	47.14		
		III	1.03–1.23			0.53	75.71		
		IV	1.23–1.94			1.24	177.14		
2100	RCP 2.6	I	– 0.10–0.48	0.80	1.93	–	–	0.10	14,29
		II	0.48–0.83			0.13	18.57		
		III	0.83–1.07			0.37	52.86		
		IV	1.07–1.93			1.23	175.71		
	RCP 4.5	I	0.28–0.85	1.17	2.70	0.15	21.43	0.47	67.14
		II	0.85–1.11			0.41	58.57		
		III	1.11–1.42			0.72	102.86		
		IV	1.42–2.70			2.00	285.71		
	RCP 8.5	I	0.90–1.48	1.73	3.01	0.78	111.43	1.03	147.14
		II	1.48–1.73			1.03	147.14		
		III	1.73–1.92			1.22	174.29		
		IV	1.92–3.01			2.31	330.00		

Q quartiles; $\overline{PMV}$ mean PMV value; PMV_{max} maximum PMV value; *Dist. PMV to* CV_{max} distance from PMV to max. Comfort value (0.70); *DI* discomfort increase; *Dist.* $\overline{PMV}$ *to* CV_{max} distance from $\overline{PMV}$ to max. Comfort value; *DIMV* discomfort increase of mean value

Table 4 Summary of sleeping thermal comfort PMV values for the case study in August

Scenario		Q	PMV ranges	$\overline{PMV}$	PMV_{max}	Dist. PMV to CV_{max}	DI (%)	Dist. $\overline{PMV}$ to CV_{max}	DIMV
2024		I	–0.43–0.35	0.59	1.73	–	–	–	–
		II	0.35–0.53			–	–		
		III	0.53–0.86			0.16	22.86		
		IV	0.86–1.73			1.03	147.14		
2050	RCP 2,6	I	–0.03–0.54	0.90	1.95	–	–	0.20	28.57
		II	0.54–0.91			0.21	30.00		
		III	0.91–1.28			0.58	82.86		
		IV	1.28–1.95			1.25	178.57		
	RCP 4,5	I	–0.09–0.66	0.99	2.43	–	–	0.29	41.43
		II	0.66–0.97			0.27	38.57		
		III	0.97–1.29			0.59	84.29		
		IV	1.29–2.43			1.73	247.14		
	RCP 8,5	I	0.13–0.83	1.09	2.28	0.13	18.57	0.39	55.71
		II	0.83–1.05			0.35	50.00		
		III	1.05–1.39			0.69	98.57		
		IV	1.39–2.28			1.58	225.71		
2100	RCP 2,6	I	–0.19–0.54	0.84	1.97	–	–	0.14	20.00
		II	0.54–0.84			0.14	20.00		
		III	0.84–1.11			0.41	58.57		
		IV	1.11–1.97			1.27	181.43		
	RCP 4,5	I	0.19–0.83	1.25	2.87	0.13	18.57	0.55	78.57
		II	0.83–1.23			0.53	75.71		
		III	1.23–1.47			0.77	110.00		
		IV	1.47–2.87			2.17	310.00		
	RCP 8,5	I	0.94–1.55	1.92	3.42	0.85	121.43	1.22	174.29
		II	1.55–1.88			1.18	168.57		
		III	1.88–2.29			1.59	227.14		
		IV	2.29–3.42			2.72	388.57		

Q quartiles; $\overline{PMV}$ mean PMV value; PMV_{max} maximum PMV value; Dist. PMV to CV_{max} distance from PMV to max. Comfort value (0.70); *DI* discomfort increase; *Dist.* $\overline{PMV}$ *to* CV_{max} distance from $\overline{PMV}$ to max. Comfort value; *DIMV* discomfort increase of mean value

8.5 scenarios, respectively. This increase means that the $\overline{\text{PMV}}$ is 14.29% (RCP 2.6), 67.14% (RCP 4.5), and 147.14% (RCP 8.5) above the comfort limit.

Regarding August, the $\overline{\text{PMV}}$ value obtained is 0.59, reaching a maximum value of 1.73 in the current situation (year 2024).

In 2050, the $\overline{\text{PMV}}$ value increases to 0.90 (RCP 2.6), 0.99 (RCP 4.5), and 1.09 (RCP 8.5). Additionally, maximum PMV_{max} values of 1.95, 2.43, and 2.28 were observed for each RCP scenario, respectively. The $\overline{\text{PMV}}$ values exceed the comfort limit by 28.57% (RCP 2.6), 41.43%(RCP 4.5), and 55.71% (RCP 8.5).

This increase continues in the year 2100, where the $\overline{\text{PMV}}$ values in the RCP 2.6, RCP 4.5, and RCP 8.5 scenarios are 0.84, 1.25, and 1.92, respectively. The PMV_{max} values obtained in each scenario are 1.97, 2.87, and 3.42, respectively. This increase results in the $\overline{\text{PMV}}$ being 20.00% (RCP 2.6), 78.57% (RCP 4.5), and 174.29% (RCP 8.5) above the comfort limit.

Therefore, from the values obtained after evaluating the sleeping thermal comfort in the dwelling, it is observed that due to climate change, thermal discomfort will increase in the coming years. The $\overline{\text{PMV}}$ values obtained in all future scenarios considered are above the maximum thermal comfort limit, reaching up to 3.5 times above this limit in the most unfavourable RCP scenario (RCP 8.5).

Considering the situations in the scenarios analysed for the same year (2050 or 2100), the results are more unfavourable as the Representative Concentration Pathway becomes more severe. Consequently, sleeping thermal discomfort increases, reaching levels 1.41–1.56 times above the maximum comfort range for July/August in the year 2050, and 2.47–2.74 times above the maximum for July/August in the year 2100, both in the most unfavourable RCP scenario (RCP 8.5).

4 Conclusions

In this study, an analysis was conducted on a dwelling located in a residential neighbourhood of Granada with the aim of determining the impact of global warming and touristification on sleeping thermal comfort. For this purpose, the effects of environmental noise pollution and the limitations on implementing natural nocturnal ventilation strategies in the dwellings were considered.

The results show that the interior temperature of the dwelling increases due to the inability to ventilate at night. According to the different scenarios evaluated, the interior temperature of the dwelling is between 4.66 and 1.17 °C higher than the exterior temperature on summer nights (between 19.36 and 4.23% higher). The greatest difference corresponds to the current situation (year 2024) and the smallest to the RCP 8.5 scenario in the year 2100 (when nocturnal outdoor temperatures will be higher).

According to the results obtained after applying the PMV model adapted to the nocturnal period, the limited implementation of natural nocturnal ventilation strategies caused by noise pollution from touristification affects the thermal comfort of the dwellings. This means that the percentage of discomfort hours in the current scenario is 33.23%, increasing to 80.32% and 100.00% in the most unfavourable RCP scenarios for the years 2050 and 2100, respectively. Regarding the mean PMV values, an increase of 94 and 239% was observed compared to the current mean value in the most unfavourable RCP scenarios for the years 2050 and 2100 in July and an increase of 84 and 225% in August.

Regarding the maximum PMV values, readings of 1.32, 1.96, and 3.01 were found in July and 1.73, 2.43, and 3.42 in August, for the current situation and the most unfavourable RCP scenarios in the years 2050 and 2100, respectively. This situation is critical in the years 2050 and 2100, as the PMV values obtained are significantly above the comfort limit.

Finally, from the research conducted and the results obtained in this study, it is possible to conclude that the inability to implement nocturnal ventilation strategies (due to noise pollution caused by touristification) along with global warming impacts sleeping thermal comfort. The percentage of discomfort hours increases in all future scenarios analysed (RCP 2.6, RCP 4.5, and RCP 8.5 in 2050 and 2100) compared to the current scenario (2024), reaching 100% discomfort hours in the most unfavourable RCP scenario in the year 2100. Consequently, the adaptive capacity of the dwelling occupants is limited, and achieving thermal comfort requires the use of air conditioning systems. This situation results in greater energy resource consumption and the consequent increase in environmental and economic impact. Alternative strategies to the current tourism model are necessary to address the challenges faced by city inhabitants, avoiding limitations on the thermal adaptation capacity of residential building occupants and promoting measures that contribute to reducing the number of discomfort hours in the current and future scenarios.

Acknowledgements María Luisa de la Hoz Torres wishes to acknowledge the support of the MICIU/AEI/10.13039/501100011033 and by European Union NextGeneration EU/PRTR under a Juan de la Cierva post-doctoral contract (JDC2022-049561-I).

References

1. Zaharna M, Guilleminault C (2010) Sleep, noise and health: review. Noise Health 12:64–69. https://doi.org/10.4103/1463-1741.63205
2. Röösli M, Mohler E, Frei P, Vienneau D (2014) Noise-related sleep disturbances: does gender matter? Noise Health 16:197–204. https://doi.org/10.4103/1463-1741.137036
3. de la Hoz-Torres ML, Aguilar AJ, Ruiz DP, Martínez-Aires MD (2024) An investigation of indoor thermal environments and thermal comfort in naturally ventilated educational buildings. J Build Eng 84:108677. https://doi.org/10.1016/J.JOBE.2024.108677
4. Roetzel A, Tsangrassoulis A, Dietrich U, Busching S (2010) A review of occupant control on natural ventilation. Renew Sustain Energy Rev 14:1001–1013. https://doi.org/10.1016/J.RSER.2009.11.005
5. Custódio DA, Ghisi E, Rupp RF (2024) Thermal comfort in university classrooms in humid subtropical climate: field study during all seasons. Build Environ 258:111644. https://doi.org/10.1016/J.BUILDENV.2024.111644

6. Jiménez Torres M, May Tzuc O, Vázquez-Torres CE, Bassam A (2024) Mapping of building energy consumption and emissions under representative concentration pathway scenarios by a geographic information system descriptive framework: case study of Mexico. Sustain Energy Plann Smart Grids: 31–51. https://doi.org/10.1016/B978-0-443-14154-6.00010-7
7. World Meteorological Organization, United Nations Environment Programme, Intergovernmental Panel on Climate Change, Global Framework for Climate Services, United in Science: High-level synthesis report of latest climate science information convened by the Science Advisory Group of the UN Climate Action Summit 2019 (2019). https://library.wmo.int/idurl/4/56814. Accessed 13 June 2024
8. European Commission, Energy Performance of Buildings Directive (2024) https://energy.ec.europa.eu/topics/energy-efficiency/energy-efficient-buildings/energy-performance-buildings-directive_en. Accessed 13 June 2024
9. May Tzuc O, Peña López G, Huchin Miss M, Andrade Durán JE, Chan González JJ, Lezama Zárraga F, Jiménez Torres M (2023) Improving thermo-energetic consumption of medical Center in Mexican hot–humid climate region: case study of San Francisco de Campeche, Mexico. Appl Sci 13:12444. https://doi.org/10.3390/APP132212444
10. Luo Y, Zhang L, Bozlar M, Liu Z, Guo H, Meggers F (2019) Active building envelope systems toward renewable and sustainable energy. Renew Sustain Energy Rev 104:470–491. https://doi.org/10.1016/J.RSER.2019.01.005
11. Sadineni SB, Madala S, Boehm RF (2011) Passive building energy savings: a review of building envelope components. Renew Sustain Energy Rev 15:3617–3631. https://doi.org/10.1016/J.RSER.2011.07.014
12. Thermal E, Nunzi J-M, El Ganaoui M, El Jouad M, Lourenço Niza I, Mendes da Luz I, Maria Bueno A, Eduardo Broday E (2022) Thermal comfort and energy efficiency: challenges, barriers, and step towards sustainability, smart cities 2022, vol 5, pp 1721–1741. https://doi.org/10.3390/SMARTCITIES5040086
13. Bienvenido-Huertas D, Rubio-Bellido C, Pérez-Fargallo A, Pulido-Arcas JA (2020) Energy saving potential in current and future world built environments based on the adaptive comfort approach. J Clean Prod 249:119306. https://doi.org/10.1016/J.JCLEPRO.2019.119306
14. Bienvenido-Huertas D, de la Hoz-Torres ML, Aguilar AJ, Tejedor B, Sánchez-García D (2023) Holistic overview of natural ventilation and mixed mode in built environment of warm climate zones and hot seasons. Build Environ 245:110942. https://doi.org/10.1016/J.BUILDENV.2023.110942
15. Bienvenido-Huertas D, Sánchez-García D, Rubio-Bellido C (2022) Influence of the RCP scenarios on the effectiveness of adaptive strategies in buildings around the world. Build Environ 208:108631. https://doi.org/10.1016/J.BUILDENV.2021.108631
16. Pereira PFC, Broday EE (2021) Determination of thermal comfort zones through comparative analysis between different characterization methods of thermally dissatisfied people. Buildings 11:320. https://doi.org/10.3390/BUILDINGS11080320
17. Delgado-Gutierrez E, Canivell J, Bienvenido-Huertas D, Hidalgo-Sánchez FM (2024) Adaptive comfort potential in different climate zones of Ecuador considering global warming. Energies 17:2017. https://doi.org/10.3390/EN17092017
18. Baba FM, Ge H, Zmeureanu R, Wang LL Optimizing overheating, lighting, and heating energy performances in Canadian school for climate change adaptation: sensitivity analysis and multi-objective optimization methodology. Build Environ 237:110336. https://doi.org/10.1016/J.BUILDENV.2023.110336
19. De Grussa Z, Andrews D, Lowry G, Newton EJ, Yiakoumetti K, Chalk A, Bush D (2019) A London residential retrofit case study: evaluating passive mitigation methods of reducing risk to overheating through the use of solar shading combined with night-time ventilation. Build Serv Eng Res Technol 40:389–408. https://doi.org/10.1177/0143624419840768/ASSET/IMAGES/LARGE/10.1177_0143624419840768-FIG9.JPEG
20. Wang Z, Cao B, Zhu Y (2021) Questionnaire survey and field investigation on sleep thermal comfort and behavioral adjustments in bedrooms of Chinese residents. Energy Build 253:111462. https://doi.org/10.1016/J.ENBUILD.2021.111462

21. Tsang TW, Mui KW, Wong LT (2021) Investigation of thermal comfort in sleeping environment and its association with sleep quality. Build Environ 187:107406. https://doi.org/10.1016/J.BUILDENV.2020.107406
22. Hillman D, Mitchell S, Streatfeild J, Burns C, Bruck D, Pezzullo L (2018) The economic cost of inadequate sleep. Sleep 41:1–13. https://doi.org/10.1093/SLEEP/ZSY083
23. Murage P, Hajat S, Sari Kovats R (2017) Effect of night-time temperatures on cause and age-specific mortality in London. Environ Epidemiol 1. https://doi.org/10.1097/EE9.0000000000000005
24. Taylor J, McLeod R, Petrou G, Hopfe C, Mavrogianni A, Castaño-Rosa R, Pelsmakers S, Lomas K (2023) Ten questions concerning residential overheating in Central and Northern Europe. Build Environ 234:110154. https://doi.org/10.1016/J.BUILDENV.2023.110154
25. Cao T, Lian Z, Ma S, Bao J (2021) Thermal comfort and sleep quality under temperature, relative humidity and illuminance in sleep environment. J Build Eng 43:102575. https://doi.org/10.1016/J.JOBE.2021.102575
26. Li S, Hu S, Liu R, Liang S, He M, Sun J (2022) Effects of ambient temperatures on sleeping thermal comfort and respiratory immunity: a winter field study in college students. J Build Eng 52:104375. https://doi.org/10.1016/J.JOBE.2022.104375
27. Castañeda JA, Vena-Oya J, Rodríguez-Molina MÁ, Martínez-Suárez R (2019) Analysis of domestic cultural tourism spend by segment in the city of Granada: an observational data approach. Tour Manag Perspect 29:18–30. https://doi.org/10.1016/J.TMP.2018.10.001
28. State Meteorological Agency (AEMET) (2024) Seasonal analysis: Granada—summer 2023. Temperature
29. ASHRAE, Guideline 14-2014 (2014) Measurement of energy, demand, and water savings. https://webstore.ansi.org/standards/ashrae/ashraeguideline142014. Accessed 14 June 2024
30. Fanger PO (1970) Thermal comfort. Analysis and applications in environmental engineering
31. Shapiro CM, Goll CC, Cohen GR, Oswald I (1984) Heat production during sleep. 56:671–677. https://doi.org/10.1152/JAPPL.1984.56.3.671
32. Lan L, Zhai ZJ, Lian Z (2018) A two-part model for evaluation of thermal neutrality for sleeping people. Build Environ 132:319–326. https://doi.org/10.1016/J.BUILDENV.2018.02.004
33. Yan S, Xiong J, Kim J, de Dear R (2022) Adapting the two-node model to evaluate sleeping thermal environments. Build Environ 222:109417. https://doi.org/10.1016/J.BUILDENV.2022.109417
34. ANSI/ASHRAE Standard 55-2023 (2023) Thermal environmental conditions for human occupancy. https://webstore.ansi.org/standards/ashrae/ansiashraestandard552023?gad_source=1&gclid=CjwKCAjw1K-zBhBIEiwAWeCOF4r3ik2ecSJAaPO1sOiq237yhFBQFi961T1Ip8WeFir5m0eYeQtMtxoCg3EQAvD_BwE. Accessed 14 June 2024
35. UNE-EN 16798-1:2020 (n.d.) Eficiencia energética de los edificios. Ve... https://www.une.org/encuentra-tu-norma/busca-tu-norma/norma?c=N0063261. Accessed 14 June 2024

The manufacturer's authorised representative in the EU is Springer Nature Customer Service Centre GmbH, Europaplatz 3, 69115 Heidelberg, Germany. If you have any concerns regarding our products, please contact ProductSafety@springernature.com

Printed and bound by CPI Group (UK) Ltd, Croydon, CR0 4YY
01/10/2026
02225958-0002